Routledge Revivals

A Philosopher with Nature

Benjamin Kidd (1858-1916), well-known for his ground-breaking application of social Darwinism in his premier work *Social Evolution* (1894), was a sociologist and a keen observer of nature. First published posthumously in 1921, *A Philosopher with Nature* is a collection of Kidd's most profound writings concerning natural habitats. Although the book is not to be considered scientific, Kidd's method of uniting biology and sociology sheds remarkable insights into the animal kingdom. This title is suitable for students interested in Anthropology, Sociology, and Natural History.

A Philosopher with Nature

Benjamin Kidd

Routledge
Taylor & Francis Group

First published in 1921
by Methuen & Co. Ltd

This edition first published in 2016 by Routledge
2 Park Square, Milton Park, Abingdon, Oxon, OX14 4RN
and by Routledge
711 Third Avenue, New York, NY 10017

Routledge is an imprint of the Taylor & Francis Group, an informa business

Publisher's Note
The publisher has gone to great lengths to ensure the quality of this reprint but points out that some imperfections in the original copies may be apparent.

Disclaimer
The publisher has made every effort to trace copyright holders and welcomes correspondence from those they have been unable to contact.

A Library of Congress record exists under LC control number: 22008377

ISBN 13: 978-1-138-18763-4 (hbk)
ISBN 13: 978-1-315-64300-7 (ebk)

A PHILOSOPHER WITH NATURE

BY

BENJAMIN KIDD

METHUEN & CO. LTD.
36 ESSEX STREET W.C.
LONDON

First Published in 1921

PREFACE

BENJAMIN KIDD was always a keen observer of nature. He was engaged more or less continuously throughout his life in carrying out systematic observations and experiments on the habits and intelligence of animals, and in the pursuit of this hobby collected a large number of careful notes which it was always his intention to publish. He died, however, without bringing this side of his life to fruition, and it has not yet been possible to publish his records and notes on this subject.

From time to time, however, throughout his life he wrote essays and articles of general interest on subjects of natural history. In the present volume a selection of these has been brought together for the first time. The first two in the book—the latest written—have not before been published. The remainder appeared in the author's lifetime, over a period of some twenty years, in periodical literature. We are able to reproduce them now by the courtesy of the journals in which they appeared.

Owing to the manner in which these essays were in the first place written they contain a certain amount of repetition which, though to a certain extent unavoidable, has been as far as possible reduced in editing the present selection. The student of Benjamin Kidd's sociological writings will find interesting side-lights upon the development of the author's mind and of the strong influence of biological studies upon his sociological work. In the literary sense some of the essays undoubtedly reach higher levels than others. In a few the artistic and dramatic genius of the author which is in evidence in his more profound works, "Social Evolution" and "The Science of Power," finds expression and renders these essays worthy of preservation on this ground alone.

The essays here collected do not pretend to be scientific contributions, being in origin written more for the author's and the reader's recreation than for purposes of advancing knowledge.

FRANKLIN KIDD.

May, 1921.

CONTENTS

A PHILOSOPHER WITH NATURE

I

WILD BIRD LIFE IN THE SEVERN ESTUARY

I

IT is low tide and early in the morning, and our boat drawing only a few inches of water is as far in as we can get, with her nose buried in the soft mud. On the right, far away to the south and looking through the morning mist like a dark bank of clouds over the horizon, stretches the steep line of the North Somerset and North Devon coast guarding the Exmoor highlands and the wild country of Blackmore's novels. In front of us is a scene to which it would be difficult to do full justice in description. High water mark is nearly three miles inland, where the low line of sand dunes rises to the skyline. Between, and stretching away on each side as far almost as the eye can reach, are mud-flats now uncovered, a great expanse of feeding ground where no human foot can travel, where no shot-gun can carry and where the wild sea-fowl find one of the greatest natural bird sanctuaries which still remain to them in Great Britain.

In this wilderness flung between sea and land we are on the borders of a country steeped in historic associations and in legends far older. Beyond the sand dunes and scarcely more than a dozen miles inland lies the site of Avalon of Arthurian legend; whither according to William of Malmesbury Joseph of Arimathea is said to have come bearing the Holy Grail, where he planted his pilgrim's staff which grew into the Holy Thorn, and where he founded the first Christian Church of Britain. It is a land where almost every site is connected not alone with history but with deeds long previous to its record, where the plains have been historic battle fields, but where the hills are moulded by pre-historic camps, or by mounds which have been places of sepulture after battle for the successive waves of invaders who came hither to take the rich land beyond before existing nations were named. Full many a heart the Danube to the Severn gave before the poet sang. Over these mud-flats Saxons and Danes, Romans and Celts, and a hundred unnamed peoples before them have sailed their keels on the flowing tide. Yet they lie before us now in the morning sun a lone expanse without mark of man on them, untamed and untoiled by any record, churning the salt tides twice daily and echoing the plaintive notes of the wild sea-fowl even as they did in the days when the fourth dynasty still reigned in Egypt.

Towards low water the tides, following ever the same channels in their retreat, have worn the mud into furrows and groins. Some are but a hand's breadth, others are wide like miniature rivers, others also are both deep and wide, for they are the mouths of inland streams which have carried their waters hither through the mud to the open sea. Following one of these natural creeks for a space, we leave the splash of the waves gradually behind us; as the boat grounds again the eye travels over the scene in search of details, while the ear in the seeming deep silence begins to pick out the sounds that reach it.

The night-feeding birds which have followed the retreating tide are still scattered upon the flats in large numbers, and the eye soon begins to distinguish the masses of black and white plumage and the specks of grey upon the brown expanse. A flock of gulls are surrounding some object in the immediate foreground and the barking, musical and goose-like notes of the larger birds come clear across the air with an indescribable suggestion of solitude and unlimited open space in the sound. A dead steer from a trans-Atlantic cattle-ship has been cast up by the waves and the birds are feeding upon it. Those of the smaller species stand in the background, only helping themselves as they may; for the larger kinds are the masters at these feasts.

We have come too near; a restless air has taken possession of the birds. A single black-backed gull resenting the intrusion sails majestically away seawards. A herring gull with pinions wide outspread circles and soars upon the breeze close overhead, directing a searching black eye of inquiry and protest upon us. The gleam of the sun on the wet surface underneath is reflected upwards upon the white plumage immaculate. There is no stain of mud; and no taint even of the recent feast upon the glorious yellow beak. These are birds in full plumage, but you notice as you look through a glass that the greater majority of the others have immature markings; for it is the breeding time and only these young birds of the previous year have this leisure. The older birds are all at the nesting haunts on the cliffs beneath the horizon.

At the line of the water's edge many different kinds of sea-fowl are congregated. The great shelducks which have followed the tide outwards stand in groups upon the higher mounds of mud, preening themselves in the morning sun after their meal, the pure black and white plumage showing strangely conspicuous against the grey background. The little merry dunlins are wading knee-deep in the water or racing on the mud; oyster catchers, feeling less at home here than on their native rocks, stand apart from the others; and mallards and shelducks rise and fall with the swell in the shallow water. Some of the latter are standing inland, resting on one leg and with beak thrust into the back feathers, but with eyes open, a picture of repose and alertness. Now a flight of ring-dotterel skims low over the water, the mass of birds swerving and curving as if it were directed by a single will, the white of the under-plumage flashing suddenly in the sun and being as instantly occulted. It is a sight which when the flocks of birds number thousands sends a ripple down the back as if one watched the evolutions of an army rendering instant obedience to a signal from the mind of a commander. Yet observing the movement now, it may be seen how the effect is produced. The passing impulse which gives rise to the sudden change of flight is often capricious and confined to few of the birds. But the rapidity of the instinctive movement of imitation by the others produces the striking effect of an army manœuvring under command.

There is a peculiar fascination in watching wild nature thus in the abandon of its native haunts and at close quarters. One of the first results that it produces is the conviction that many of the currently received theories of the origin of language will be revised when we are wiser. The most primitive language is undoubtedly a language of the emotions. But the language of emotions is not, as might be expected, confined only to members of the same species; it is amongst birds, at least, a kind of *lingua franca* understood even by widely different species. When one has lived under other conditions with some of the wild birds here seen in their native surroundings it comes with a certain surprise to observe how the signs and sounds with which one has been familiar elsewhere are interpreted in their wild haunts by their own kind and by other birds for values which are evidently well understood. It is the breeding season. The eye lingers on the actions of a sheldrake standing before his mate with other birds of his kind in the background. The excited pump-handled movement of the head and neck is accompanied by a continued protesting and haranguing series of notes which has evidently its exact emotional significance. Yet you become conscious that the declamation possesses depths of meaning even like the song of a nightingale. The emotion rises and falls until the scene reminds you irresistibly of the declamations of the South African negroes as you have seen them under the influence of native narcotics, when it was the sounds and not the words which conveyed to the listeners the intense emotion of the speaker.

Now there comes up the wind a flight of mallards. These wild ancestors of all our domestic ducks lower themselves into the pool from their flight, cutting the surface with the action of a swift boat taking the water. They have been feeding inland, and the sight of this sunny sheltered backwater soon produces a remarkable effect. A preliminary chatter and a single bird seems to go suddenly mad. With half-outstretched wings and with lightning-like rapidity it takes short glancing dives beneath the surface. The chatter is taken up by the other birds, and the infection spreads instantly. Within a few seconds every duck is darting through the water, under or over it, as if bewitched. The excitement is communicated to the swimming birds of other kinds standing near and within a brief space one kind after another, each keeping apart by itself and threatening or protesting to the others, joins with wild cries in the boisterous scene in the water. It is all play. Yet the signs and cries which accompany the wild movements are evidently as clearly interpreted by the various kinds of birds as if the language had been spoken words and as if the scene and actors had been exclusively human. Now at a low raucous note from a single bird all action is instantly frozen and every neck erect. From the outskirts of the crowd comes the quick plaintive call of the warning curlew; and every bird, thrilled by some primordial instinct of alarm, is instantly in the air with a roaring sound of wings. It has been a scene of life excited, full of understanding, eloquent of communicated feeling. There is indeed, one feels, a language of the emotions among animals. Yet it is not, as it is often absurdly imagined, a language of words. It is a language of sounds and sometimes of signs. But the sounds do not represent words. They are thrills and utterances which reach the depths of primitive emotion. They are declamations, intonations, cadences, incantations. And beyond doubt they are capable of powerfully and instantly reproducing corresponding states and shades of intense feeling in those affected by them.

The tide has turned and has now begun to flow, the water rapidly lapping its way over the mud and singing on the half-dried surface as it recovers it. Remote in the distance a kestrel hangs in mid-air over the sand dunes, looking, save for the difference of size, curiously like the turkey-buzzard as it may often be seen on the wing in Southern California. Yet not, like the latter, on the look-out for carrion, but with an alert eye watching the small birds in the brambles below and ready at an opportune instant to swoop on its prey like a bolt from above. Nearer still but higher in the air a large bird, long and slim in body and strong of flight, is making for the south-west, looking almost headless as it wheels in the air. It is the peregrine falcon of lordly fame in the spacious days of hawking. The bird still nests near by, and with the single exception of the raven it is the master of all that flies here. And yonder, travelling high over the inland marshes with its large wings flapping in slow and stately progress, is its ancient and noble quarry the heron. Many of the old heronries still exist. There are two not far distant and the birds come here daily to fish in the marshes.

As the tide advances it is possible to land and make a long detour to reach the belt of shingle which runs east and west at the line which marks the reach of the highest tides. A solitary ring-dotterel or ring-plover, and now another, runs with suspicious motions which avoid notice across the ribbon of sand toward the sea. It is the nesting season. Above high-water mark the birds have been scooping shallow depressions among the small pebbles. These are the " cock " nests which precede and accompany the real ones, and the eye searches the beach closely for the characteristic eggs.

On this solitary coast, where one looks out along the fifty-first parallel toward the Western Continent across many of the main ocean highways of the world, the beach beyond reach of the highest waves is covered with undisturbed mounds of the flotsam and jetsam of sea traffic. The buried cities of the world have left us relics of the ages of man. But what a record of the present civilization of the world these heaps would yield to the followers of some post-historic Schliemann if they were to be suddenly entombed and to give up their secrets again to a distant age ! Floated fragments of every kind, drift-wood and bark, cinders, seaweed, and the white stones of the beach, are intermingled with the offal of ships and fleets, the droppings of ocean-liners and tramps, of fishing craft and the floating fortresses of war. There are relics of divers nationalities, of many kinds of human products, of universal art and literature and of most human customs. It is the story of a world of labour and sweat and the silent tyranny of things that are strong; with here and there a grimmer relic of the deep speaking of some unrecorded tragedy of the sea. But everywhere corks, corks, corks. Thousands and millions of them. Most of them worn and fretted by the waves. Some new and familiar; some with the marks of the lordly vintages of France still stamped upon them; some evidently cast away in distant latitudes and longitudes, bearing strange devices and legends in unknown tongues; but all borne here by the sea. There have been ages of stone, and of metal, and of the potter's art. But few of us realize that we are ourselves living in the most characteristic age of all—the Great Bottle Age; the age when universal man drank things out of bottles and strewed the earth with the shards thereof and the ocean itself with the corks.

And here in the pebbles amid all these disjected sweepings of the world and just above high-water mark, the little ring-dotterel still places her nest, even as she did before man moved on the waters or troubled the earth by going up and down in it. Only a slight depression is scooped in the grey pebbles and coarse sand. The grey-yellow and spotted eggs, which are four in number, lie with their small ends together. They look like emblems of a peace enfolding all things as they lie here in the warm sun. Yet do they too bear the marks of the world-old stress upon them. For they are so protectively coloured to their surroundings that they are almost invisible at a short distance. Thus do the ages of stress overlap each other and ever with the same meaning in them.

The lapwings are tumbling and crying over the

marsh pastures. *Pee-wit, pee-oi-ku-si, pee-wit,* their plaintive musical calls resound through the air overhead. The nesting season is well advanced and the birds rise one by one out of the herbage with subdued excitement visible in their movements. These grass-covered flats which extend for miles have all been reclaimed from the sea, and much of the salvage would again be retaken at high tides but for the system of banks and dykes. They are a favourite breeding ground for great flocks of the lapwing, as the common plover is generally called.

The birds place their nests in fixed relationship to the lie of the ground, desiring above all things to be able to get off the eggs and slip away unobserved by taking advantage of some neighbouring hollow or depression. Without knowing this one might look here for hours and not find a nest; and yet with such local lore half a dozen or more are discovered in almost as many minutes. The eggs lie in little round depressions in the ground with scarcely any attempt at providing lining materials. They are no larger than pigeons' eggs, and they lie always four in a nest with the small ends together. If you would have it otherwise and change the position and return you will find that the birds have altered your handiwork. The ground colour is grey with a blend of yellow and green in it, and it is thickly splashed with large and small blotches of black. On the rough ground and in a bright light and surrounded with grass they are almost invisible at a short distance. These are the plovers' eggs of fame which fetch such high prices in London for consumption at fashionable weddings. As they lie here amid the lush herbage starred with flowers, with the gleam of water everywhere around them, and with the open air and bright sunshine and blue sky above them, they seem indeed no unfit emblem of hope.

And the wanton lapwing himself? There he stands in the distance, with an anxious eye turned upon us; poised on one leg with the other half lifted; now bending forward his body gracefully, now breaking into a quick run with his plumage showing the shimmering of green amid the black and pure white in the sunlight. And now at last turning to us his beautiful crest.

In the spring the wanton lapwing gets himself another crest. Well did poet immortalize him and link him with other vernal emblems of the season in a passage which here in particular one feels to breathe the spirit of refulgent life as it glows ardent and radiant in the increasing procession of our northern year.

Some of the birds come quite close with strange and anxious antics, both in the air and on the ground, as if inviting us to follow them. The practised eye reads the situation. The young, which leave the nest as soon as they are hatched, must be somewhere close by. You look and stare; there can be nothing, for there is absolutely no cover. Yet even as the eye relinquishes its quest something stirs; and you see it is a young lapwing among the clods, just hatched, which with its protective colouring has been hiding itself with extraordinary effectiveness against the lap of mother earth. You take the quaint little piebald ball of fluff in your hand and it stands erect, looking, as very young animals so often do, wizened and aged after the tremendous experience of an hour's independent existence.

In the long grass where the furrow slopes steeply and the ground is dry it is a delight to fling the body at full length. The swallows fly high overhead in the still air. The harsh laugh of the green woodpecker comes from far inland. The whistling wings of the wild ducks sound above in the air as the birds change their feeding grounds with the tide. They are nearly all males to be seen here at this season, for the females are sitting elsewhere and their partners have to shift by themselves. A large dark bird with steady and powerful beats of its black wings is making for the hills in the east. It is the raven, now far less commonly seen here than formerly, but the bird of superstition which shared the sea journeyings of the Norsemen when they came here still lingers on this coast despite persecution.

As the eye descends it rests on something brown quite close in the thick grass. As the herbage is pressed down gently to bring it into view two little hares are exposed to sight, lying as close as possible together head to tail. There is not the slightest movement from them ; even when one stands up and walks round them they stir not. The ears are laid back flat against the body and only the just perceptible motion of the beautiful brown eyes, exposing at times the faintest rim of white at the edges, shows that every sense is alert. These little creatures form the easy quarry of many birds of prey overhead. They know instinctively that the slightest movement is revealing, often sealing their doom instantly. Hunted creatures are they from birth to death. And yet when tamed, and you speak from experience, full to overflowing of the frolic and wanton of life.

The shallow creeks run far into the grass country, and along the edges of the tidal water the redshank sandpiper loves to place her nest. Where the *rhines* holding the surface water broaden out the sides are fringed with rushes and young sedge. As you look a moor-hen with the movements peculiar to all the rails sails shyly across the water, jerking her body as she swims just like her Virginian relative. From her movements it is easy to discover her secret. The nest, with nine buff eggs spotted with reddish brown, is in a tuft of rushes by the bank where the cattle feed almost to the edge. Further on in a surface-pool another nest of the true water rail, with six eggs, lies in the rushes on the brink, the billows of rich grass flecked thickly with flowers surrounding it.

On the top of the embankment which divides the saltings covered by the tides from the marshes beyond a black and white object attracts attention. It has evidently recently come there. As you advance toward it you are conscious of a slight shock. It is a pair of large black and white wings. You lift one of them and find that it is connected with the other by the whole framework of a bird plucked of the flesh and with some of the fresh fragments of red meat still hanging to the bones. The wings are those of a lapwing certainly alive not more than an hour or two ago. Scarcely a feather has been disturbed. The beautiful green and black and white plumage still gleams in the sunshine almost like that of some tropical parrot. It is the recent kill of the peregrine falcon. You passed a similar pair of wings belonging to a large sea-bird on the beach and just now another pair of pigeon's wings on the grass. They have all had the same history.

In the midst of this world of beauty thus is nature red with rapine. Yet would it be the profoundest of mistakes to endeavour to read the emotions appropriate to a different plane of being into these incidents of the universal stress of nature. For to this stress the whole of this glowing world of fitness and potency is undoubtedly allied. It is the product, born of it and in it. To suspend the stress might indeed be to take from life something of the sting of its pain. But it would certainly also be to take from it that constituent of its deepest pleasures which passeth understanding—the joy of a world of fitness functioning in achievement. Even to humanize the stress is not to suspend it. Far otherwise. The secret of ripening humanity is indeed nothing else than the secret of this higher fitness; the fitness of apparent failure; to be able not only to succeed consummately but to fail infinitely for others.

The tide creeps quickly over the lower fringes of the saltings. On this no-man's land the life of the ocean struggles with that of the land and here the sea-fowl and the land-fowl meet. The surface is covered with salt herbage, on which the sheep thrive, cropping it between the tides, and it is hollowed in places into long irregular pools. These hold the sea-water and, surrounded by the prevailing dark grey tints, they reflect the sunlight as if they had been pools of white molten metal. The mud at the bottom is marked with the feet of many birds. Some pools have dried up, leaving the markings legible and the footprints of the land-birds, rooks, carrion crows and jackdaws, are seen mingled with the prints of the webbed feet of the sea-fowl. Here a pool which has run dry has been missed and it reveals at a glance the kind of harvest the birds have gathered. On the damp surface just at the lowest part of the bottom there lies a small heap of sea shrimps, mostly dead but with some of the survivors still gasping in the hot sun. The high spring tides brought the shrimps here but the treacherous sea has gone and so left them. We speak of the unerring instinct of nature, but often, as in this case, it is indeed no more unerring than the chequered wisdom of man.

With the incoming tide the sea-birds move from the beach to higher quarters, and where the long line of the Mendips drops at last into the sea the slopes of the cliff are dotted with white plumage. It is a steep climb upwards. The springy turf which clothes the high ground is close-cropped and the little rock-rose, a botanical survival almost peculiar in Britain to this headland, stars the green in places. The view opens over the water till it includes, far in the mist on the horizon, the spectral-seeming ships on the sea roads. They might indeed be phantom ships of another world for all the relation they have to this. Thus have they passed day by day, even since Cabot set his prow to the setting sun on these waters and discovered a western continent; and thus have they passed long before him. But the wilderness is still the wilderness here. In some of the almost inaccessible slopes the rabbits find safe refuge and their white cotton tails twinkle everywhere as they scud to their holes at the approach of an intruder.

This is the nesting-ground of the great shelducks, the characteristic sea-fowl of the mud-flats of the Severn Estuary. The birds circle round uneasily in the air. Some stand alert on the ground, well out of range, with their long necks erect but with the black heads always in motion. When the birds think they are safe the curious habits of these sea-fowl may be watched from a distance. They waddle amongst the rabbit burrows quite at home, ever and anon disappearing in the dense growth of low bushes. They are on the best of terms with the rabbits, and the association is curiously suggestive of the owls and prairie dogs as one sees them on the western plateaus of America two days out from Chicago. The birds rise when they are approached and with the hold of some primitive instinct strong upon one the rough intervening ground is soon covered. Peering among the ferns and thick undergrowth, scratched with the brambles, hot in face and daubed with the red earth, you are rewarded at last. A roughly made nest full of large eggs, much larger than those of the common wild duck and approaching in size those of the wild goose, lies in the mouth of a rabbit burrow in the dim light under the dense bushes.

The sight has an indescribable effect upon a range of latent emotions. One watches the unloading of tons of bullion bricks from an ocean liner without the stirring of a pulse, and even with a sense of the uninteresting squalor of the scene. But this nest full of large creamy-white eggs, stained with the red earth, revealing the last, inmost, anxious privacy of wild nature in its secret haunts, what primordial depths it uncovers! How one turns again glowing and transformed, the hunter, the savage, the utterly unknown man, with the gorgeous sense of achievement holding him by the throat.

What an incomprehensible world it is! Are these things of the realities of life? Nay, rather are they not after all the ultimate realities: the emotions which, pent at last in the machinery of social systems and civilizations, in the conflicts of national history, in the adventures of financial strife, even in the daring quests of science after the secrets of the world on lonely mountain tops and in the innermost recesses of the laboratory, lead men ever on to the same goal—the last exulting sense of self realizing itself in achievement?

There are last things as well as first things. Yonder where the wild sea-fowl circle and scream, the day falls towards the only sign of man's handiwork here—the dismantled fort, silent and obsolete, at the end of the promontory. For the devices of war have changed, and with new knowledge have come other inventions which render the purpose of this battlement vain. In a few hours as the sun sinks it will be what the poet saw in imagination, a looming bastion fringed with fire; but empty, useless, abandoned. Thus it is that science ever from hour to hour condemns her own handiwork. It is the things of nature alone which are eternal.

2

II

WILD BIRD LIFE IN THE SEVERN ESTUARY

II

OUR boat is on the flood tide with the wind behind us. There are few places in the world where the tides rise higher or advance more rapidly than in this estuary where the impulse from the Atlantic received in a mouth fifty miles wide is gradually compressed between long narrowing shores as it ascends inland. The leagues of mud flats, impassable by human foot, which at low water stretch in all directions on the Somerset coast, are the safe retreat and feeding grounds of great multitudes of wading seafowl. Yet these are scarcely more of a natural sanctuary for one class of birds than are the wide ranges of the marsh country beyond them for another—country all reclaimed from the sea and at many points still below its level. In this flat land the tidal waters flood the surface-leads and river mouths far inland. A few hours ago our boat fell seaward on a slight river at the bottom of a trough, a ribbon of water with the blue sky above and the sloping banks of mud ascending to the skyline from either side.

But now it is borne rapidly in the opposite direction, on a heaving breast of waters, with a view stretching far to the horizon on both sides and with pulses which have drawn from out the ocean bounding under our keel.

As we ascend into the land on a tide almost at the flood the long lush grass at the full sides streams in the salt sea-water. One after another we pass the mouths of shallow creeks which bifurcate into the marsh country. Entering one of these and following it for some distance the boat drifts in the sunshine level with a waving sea of flowers and herbage, out of which the tumbling and crying lapwings rise. Almost at one's elbow one of the birds furtively takes wing, and its nest with the eggs chipped, in the last stage of incubation, is plainly visible from the water. A little further a beautiful and graceful grey-brown bird with long legs showing bright orange-red rises. It is the redshank, a bird that loves to sit close to water, and the nest with four pear-shaped buff eggs flecked with dark brown is in the grass almost at the edge and not a dozen yards from that of the plover. The boat brings to where the green surface, studded with cowslips, reaches to the tide. It rocks gently as it rests, and the soft swish of the water as it rises and falls amongst the grass sounds almost like regular breathing. It is the pulse of the far distant Atlantic losing itself here at last among the summer herbage.

We are in the ancient land of Damnonia, the local kingdom of that name of the Britons before the coming of the English. It is the country of King Arthur and his knights of story and legend. It represents a corner of the earth where the Celt has struggled for ages with his compeers, where he has met the Roman, and mingled with the Saxon and his kindred, and left a rich compost which has wonderfully fermented and fertilized the world. But the tide of history has flowed round rather than over these plains. Yonder on the horizon where the limestone hills climb upward, the Romans came to the lead mines and the road ran west and east to meet the Roman fosse which crossed the country to the ancient Aquæ Solis—the modern city of Bath. On the slopes of the hills the Roman villas rose and flourished. But when the soldiers of the second legion under the Emperor Claudius looked out from the heights over this country they saw only a swamp and the waters of an inland lake with the Tor which is now Glastonbury rising at one end. The inland lake has gone, and the swamp has been partly reclaimed. But it has become a land of water-courses overgrown with tall bushes, and deep *rhines*, which carry the drainage to the sea. It is for this reason a country in which wild nature has remained in large part unchanged for centuries.

In the still warm air the rooks sail overhead carrying in their throats to their young the food which they have gathered in the open country. The distant cawing of the birds at the nests in the elms round the hamlets far at the foot of the hills just reaches the ears and suggests an infinite tranquillity. In the nearer stillness the subdued *krack*, *krack* of a moor-hen in one of the water-leads comes on the air mingled with the twittering notes of the swallows as they skim the surface. Down the wind reaches the sound of whistling wings, and a flight of wild ducks, all males—for the females are occupied elsewhere—passes in the middle distance, the beautiful plumage of the birds with the white neck-circles showing plainly in the sunshine. And from far overhead, from above the crying plovers, falls the song of the invisible sky-lark—Shelley's "blithe spirit"—dropping its cascade of notes from the blue heaven. Another song comes from a second bird in a different quarter of the sky. And yet another from a third, the notes mellowed almost to stillness in the distance.

As the sounds of the landscape mingle with the faint but all-pervading and indescribable odour of growing herbage and young leaves, and the scent of the early hawthorn and the late cowslips, one feels on the brink of one of the secrets which primitive man probably shared with wild nature, the secret which is still presented to us in the unfathomed mystery of the migration of birds and wild creatures. For these elusive scents and sounds hold one by the throat and bring up to the surface of consciousness by association a hidden world of the most powerful emotions. A native of this land, a man of education and culture, landing at a neighbouring port after many years' absence and going at once into the country on such a morning as this amidst the growing herbage and flowers was found rolling himself on the ground in the smell of his native fields like a wild animal. One realizes thus how the call of the wilderness or the desert reaches men pent in cities in something on the spring air, or on the autumn wind, and overmasters them and commands them. And so doubtless it is with wild birds and wild nature in the uncontrollable impulses of migration.

In this country, chequered by watercourses, and where the game-keeper comes not, the balance of nature is preserved as of old. A black and white bird, as large as a pigeon and with long graceful tail, chattering as it flies, alights on the ground some distance off. It is followed by its mate, and you see they are both in anxious attendance on seven quaint-looking young ones just from the nest and as yet almost tailless. It is a family of magpies. This bird which has its place so firmly established in the folk-lore and literature of European peoples has become extinct in many parts of the country because of the persecution to which it is subjected by game-preservers. There are many nests here, and in the spring-time the dome-shaped structure silhouetted against the sky in the low trees is a characteristic feature of the landscape. Yonder in the topmost branches of a low ash sits the solitary carrion crow. He also has lost character elsewhere, but he finds a refuge in this land, and the nest, always built alone, is in keeping with the hunted habits of the bird.

In one of the distant water-leads the eye catches a grey object against the background of green, looking strangely foreign to the landscape. As you advance cautiously it proves to be a large blue-grey bird standing in the shallow water. It is a striking sight at close quarters, with something quite eastern in the appearance. The tall legs lift the body high above the surface. The flexed neck is tucked into the shoulders. The long murderous looking beak is poised downwards as the bird stands in an attitude of rapt attention. It is the heron watching for its prey in the water, the most beautiful bird of Western Europe.

Save for the difference in length of the body, it might be the great blue heron of the United States, so similar is the appearance. It is the same general look of the plumage, the same stoop of the shoulders, the same trick of attitude, the same poise of the same bayonet-shaped bill. We marvel at the variability of life. Yet it is surpassed by the still more astonishing conservatism. One is startled when meeting the second of two brothers who have lived in different countries, and who have themselves never met, to find that they have developed each the same mannerism of stroking the side of the nose in the midst of an argument. But these little identical tricks and mannerisms of biological relatives who have never met and who have been separated in their careers by vast intervals of geographical space and geological time, are more startling. They serve to reveal to us as by a flash not only the profound complexity, but the almost incredible stability of the matter which constitutes the physical basis of life.

As we travel inland on foot the scene changes. The surface of rich land formed from mud covered by the sea at a previous time gives place to a layer of peat marking the site of the swamp and lake of ancient days. Many parts are still covered with water and are overgrown with deep sedge. In other parts the heather has nearly extinguished the rival vegetation, while in others still the rich meadows march with the wilderness. This land is the retreat of vast numbers of water-fowl in winter

—wild geese, mallards, water-rails, teal, curlews, coots, snipe, bitterns, moor-hens, and various kinds of plover, and many remain to breed.

The wild scene which opens up to the sky on every side suggests the past and the history of the past at almost every point on which the eye rests. Yet one walks far through this country in the early summer noonday without encountering any human creature. It might be nature in her primeval mood, so silent is the landscape. In the prevailing stillness one becomes gradually conscious of only one sound which seems to haunt the footsteps. The cause of it must be in the distant wood in front, but there is no one when you arrive. It must be in the open space beyond and you expect to see figures in the fields and busy men at work, but you emerge again and still there is no one. The sound is as difficult to define as it is to localize. It suggests now the hum of machinery or again the distant bleating of goats, or yet again the subdued converse of people close at hand at work. But there is never any one, and it remains a whispering sound always about one in the air.

A snipe, uttering its sharp *tscaaap*, *tscaap*, rises from the heather, and now another, and you look long in the coarse grass for a nest. The carrion crow has been busy, and the traces of broken shells strewn in the marsh mark his work. At length a nest is found containing eggs. It is the slightest of structures, a few grass stalks in a dry depression in the ground. The four beautiful, almost invisible eggs, like all others in the open, bear the imprint of their surroundings in their protective colouring. Olive-tinted they are, but with unusually large peat-coloured blotches which almost run together on the surface. Another bird rises from the marsh, and the eye follows it upward in the air. It ascends till it reaches a height of about a thousand feet, when it begins a series of peculiar evolutions, the bird descending rapidly in a curve and again mounting to repeat the action. As the eye searches the sky several birds, all snipe, are seen in other directions, each engaged in similar evolutions. Suddenly the cause of the peculiar ghostly sound in the landscape is revealed. It comes from these birds high in the air. It is the bleating or drumming note of the snipe in the breeding season. The sound is emitted by each bird at the moment of its downward course through the air, and it ceases immediately the lowest point of the curve is reached. It is supposed—as the result of experiments—to be produced by the vibration of the inner web of the outermost tail feathers of the bird, as it makes its descent.

In these various ways among animals of expressing the intense emotions of the mating season we catch a glimpse, almost as if we looked down the corridors of time, of the infinite possibilities that have always been latent in life. We imagine the complexity of language and we think then of the mechanism of voice among the higher animals, and immediately we conceive it as if it were the sole means for communicating by sound emotion from one creature to another. Yet with what a range of instruments have the sounds and emotions of the love moods of life been in reality communicated. From the shrillings of the cicadas, or the scrapings of the crickets, or the lonely ecstatic lovetappings of the death—watches, to the thrills of the bob-o'-link—from the tail features of the snipe to the throat of the nightingale. The higher animals are built on the simple and effective lines that have won out in the struggle for existence. But the fact is that life would have honoured any draft whatever on her, even, as she has proved, to gills on our fingers or lungs on our toes—according to circumstances. Our ears are in our heads indeed, but life would not have boggled at placing them anywhere, even in our legs, as in the grasshoppers.

East and west through the shaking land run the narrow raised ways which mark the original level of the country. The surface on either side has long since been cut away and these strips of high ground which have been left have served as primitive roads over which the peat harvest in former days has been carried away. Bunyan must have seen a road like this, for, as in the way through the Valley of the Shadow there is on either side a deep ditch—here filled with black water—or a dangerous quag. But on the narrow ground between it is dry, with flowers everywhere : while the crisp smell of the moor, and of the antiseptic peat, lingers on the nostrils like the pungent aroma of the pepper trees and the blue gums in Southern California.

Overhead in the air the wild ducks circle—still all males. Thick on the air, travelling ever in one direction, come the feathered reed-seeds of last year. On all hands stretch the meres and leads of dark water. The deeper parts are thick with tall reeds of last year, showing at this season large heads of white down scattering on the wind. The shallows are overgrown with long sedgy grass, rising in places into high billows or still higher rounded tussocks. The firm ground between is covered with heather and low birch trees. And everywhere, and between everything, shining water. We have reached the breeding-ground of the mallards.

It is but a moment to make preparation for wading. There is no bird that swims or flies which is capable of exciting so persistent an interest in the secrets of her life as the mallard. In alertness and shyness, in craftiness in placing her eggs, in devotion to her young, and in the extraordinary tricks of avoiding pursuit which both parent and young have developed, the wild ancestor of our breeds of domestic ducks has few equals in the wild; and if blood be the price of efficiency, beyond doubt she has paid in full: for war from times primeval has man waged on her for her eggs and succulent flesh. As we advance through the marsh a scene of disquiet spreads in front. A few ducks have joined the drakes circling in the air. The moor-hens croak in the water-leads. A water-rail's nest resting in the water but daintily woven in the reeds and containing eight eggs is passed. Not thus lightly does the coot build in the water close by here a nest which is always founded on a submerged heap, laboriously gathered, of last year's sedge. Emerging at last on the dry heather and wading knee-deep through it a small duck-like bird which disappears with rapid flight is disturbed. The nest, containing nine creamy-white eggs almost concealed in feathers, is in the thick cover. But it is a teal's and there is still no trace of the mallard.

Where the water is ankle-deep in the long marsh grass strewn with last year's reed-stalks something at length catches the eye. When the sedge-bents, which stream over it and serve to screen the sitting bird, are pushed aside a mallard's nest with eggs is disclosed to view. The yellow-blue eggs are quite warm and the mother-duck has evidently only just left them. In the thick sedge-grass with water still over the ankles you pause later with back to the sun to watch the birds circling uneasily overhead. Lifting a foot to advance again, the marsh seems suddenly to explode at the spot on which you intended to put it down, and a dark mass lifted an instant in the air falls again in front. It is a second or two before you realize that the object is only a brown duck quacking loudly and wildly flapping an injured wing.

Instantly as the eye gets back to the spot from which the bird has risen the cause is revealed. It is a sight which makes one feel like a bungler and intruder upon the privacies of life. No wonder the mother duck all but allowed herself to be trodden on. She has been sitting on a nest full of little ones just emerging from the shell.

All the little ducks save three have freed themselves from the shells, and some are already so active and so ready to scamper out of the nest that they have to be restrained by hand. But as the mother, still beating her apparently broken wing, passes out of eye-shot quiet is gradually restored. You have heard from the beginning of the instinctive fear of young wild animals for man. But what a libel it proves to be on nature when taken thus at the font. You have read that the young of the mallard, when hatched out with tame ducks by a foster-mother, are inherently wild and intractable. Although you have tried it again and again with eggs from the wild and have found no fact to justify such imaginings, yet are you still scarcely prepared for what follows.

The little ducks missing the cover of the mother come out of the nest into the sedge and shallow water. They find one's bare feet as one stands urgent that the camera should arrive and, without the slightest instinctive fear, begin to nestle on them for warmth, one and another turning a comical and intelligent little black eye upwards, as if with nascent wonder at the size and aloofness of this parent. You wonder how long the wild duck has been here. No doubt the hosts of King Alfred, when he hid in these marshes from the Norsemen a thousand years ago, found her here. No doubt the soldiers of Claudius long before him flushed her when they came. Probably even in the days when the woolly rhinoceros left its remains with those of the cave-man in the hills yonder, she was here. During all this time she has probably been the most universally hunted creature on earth. And the spent cartridges of the modern man strew the bog around you. Yet here are these little creatures on your feet. You take one of them in your hand, and this heir of the ages of the blood-feud shows no fear of you, even tilting its little beak to look inquiringly in your face; evidently thinking no evil, to all appearance hoping all things and believing all things, but certainly quite willing to take you on your merits for good or evil entirely without prejudice.

You put the little creature down in deep thought and pass on. Looking back, the mother bird has alighted on a tussock near by, and the more active little ones are streaming out of the sedge to her. She is chattering with emotion, every feather quivering with excitement. The hold of the Great Terror of Man is upon her. In a few days, nay, in a few hours, she will have taught it to them and they will have passed irrevocably into another world. And yet you saw the little ducks. They knew nothing of it.

Oh, you wise men who would reconstruct the world. Give us the young. Give us the young. Do what you like with the world, only give us the young. It is the dreams which they dream, the Utopias which they conceive, the thoughts which they think, which will build the world. Give us the young before the evil past has claimed them, and we will create a new heaven and a new earth.

The afternoon shadows fall with lengthening lines on the black ground as we advance up the valley. Here the peat cutters have been at work, and the deep brown-black of the bare surface absorbs the light and gives a sombre effect to the landscape. The lines of freshly cut peat stretch away to the distance with water gleaming between them. The latest cut blocks look like huge slabs of moist black cheese, and are laid nearly flat. The dry ones, shrunken to a third of the size, are piled in heaps which in the last stage of all are in size and shape almost like hayricks. Where the ground rises and the long ferns grow beneath the trees a bird the size of a small dove, mottled brown on the back, but marked like an owl or cuckoo on the under plumage, lies dead on the ground. It is a nightjar, one of the last summer migrants to arrive, and the neck is torn where it has struck in flight against overhead telegraph wires. One recognizes the night hunter at a glance. The profession of these insect-feeders who have abandoned the struggle for existence in the day and taken to hawking for moths in the night must be a successful one, for the dispersion of the birds in the present age of the world is almost universal. The general resemblance of the bird to the American night-hawk and to the whip-poor-will is immediately apparent. The enormously wide gape with the mouth fringed with bristles and coated inside with a sticky secretion is noticeable.

A rail swims across one of the pits of inky water, jerking her tail with exactly the same little mannerisms which one sees in her relatives in other lands. Through the reeds in the further distance the long neck and motionless grey head of a solitary heron watching the intruding footsteps is just visible. The bird in this attitude resembles the stork as one sees it fishing in the reed-marshes along the Rhine; but it does not stay to be approached, taking flight immediately, the gaunt legs straggling behind as it rises in the air; while the long neck, at first outstretched, is tucked rapidly into the shoulders.

Low down across the sky comes a bird which looks like a pigeon. Yet it still more closely resembles a hawk. It is being followed and mobbed by small birds and the grey plumage is seen to be barred like a hawk's, as the bird comes to rest in the topmost branches of the thick clump of high bushes on the right. Suddenly there rings out from the bough, clear, soft, and penetrating in the stillness, the most characteristic bird-note of early summer throughout Europe—the double note of the cuckoo.

The bird is easily visible through the leaves as it sits for a moment repeating the well-known call from which it takes its name. A certain fascination attaches to every movement of the cuckoo seen thus at close quarters. The vast wanderings of the bird during the year in two hemispheres; the shy and solitary habits; the sudden return in spring from out of the unknown, uttering as it comes the mating cry which resounds everywhere over the plains and woods, the mountains and wastes of a continent; the remarkable instincts of a creature whose males greatly outnumber the females; and above all the parasitic habit which has rendered every cuckoo at the beginning of its life the central figure in a tragedy, the details of which while they run counter to the strongest instincts of human nature exceed in grim actuality any possible description of them, have so fixed the cuckoo in the imagination of European peoples that it has left its mark indelibly impressed on their languages and folk-lore.

The extraordinary restlessness of the bird is apparent. It moves through the branches and thick foliage still uttering its call, for now is the full noon of the mating season; but still also searching for food, for always is the cuckoo hungry. It flits now to a bare stump, and with a pocket-glass you catch a full view of the bird so rarely seen at close quarters. It is a beautiful creature, the glorified and perfected image of the young bird of an earlier stage; for the young cuckoo in its browner immature plumage has shared many a midnight vigil with you as the long nights of our northern winter have closed down upon it in captivity.

The long tail hangs gracefully down; the large projecting wings reveal the immense powers of flight; the short legs and the look and general poise of the bird suggest to the eye even of the artist some hidden kinship, now with the swifts or night-jars of the Eastern hemisphere, and now again with the insect-feeding night-hawks and whip-poor-wills of the Western. The plumage and actions of the bird on the other hand stir something deep in the mind which associates the cuckoo with the birds of prey and with the owls in particular. But if the mind lingers thereon the weak claws immediately rule the bird out of all such categories.

But the beak, the graceful, slender, slightly curved beak. Even a tyro in the knowledge of evolution recognizes the significance of the problems which it suggests. Thereby is the bird evidently placed apart from all the birds of prey which have solved the problem of existence by tearing and feeding on flesh. Thereby is it apart from birds like the parrots who with their strong beaks have solved the problem of the mastery of the fruits of the earth. Thereby is it apart from the foraging families of the crows all the world over. And thereby is it placed apart also from the immense army of birds who have solved the problem of feeding on hard seeds by grinding them afterwards in their gizzards. A bird, large almost as a pigeon you see: and yet with the beak of the innumerable small species which hunt the succulent caterpillar among the green leaves of the world.

This indeed, taken in connexion with the nature of its range in the Old World, is the problem of the European cuckoo. One sees how the bird stands related to many orders, but related to them at the parting of the ways from whence evolution has carried life to new horizons. In the solution of the problem of feeding itself the cuckoo has simply followed the sun north and south of the Equator. But in the solution of the problem of feeding its young it has been masterful, sinister, original.

The bird flies down rapidly into the open space close to some low gorse bushes from whence a peculiar gurgling sound seems to come. As you reach the spot with some difficulty through the long ferns two cuckoos rise from the surface. It is in one of these open spaces on another occasion that you have found the bird in the act of laying upon the ground and have picked up the egg still warm. For the cuckoo does not usually lay her egg in the nest of her victim, but carries it thither in her bill. The selected victim in these parts is generally the little hedge-sparrow, which feeds its young exclusively on soft insect food.

In a few weeks more the tragedy in which the young cuckoo figures will be in full progress in the hedges here. If you had not personally witnessed every detail of it, it would be almost impossible to believe fully in the concentrated purposeness of the long series of acts by which the end is attained. The young cuckoo after it is hatched out by the foster-parent lies for about two days helpless at the bottom of the nest. Then the spirit of its ancestry descends upon it. The young of the parasitic American cowbird is usually content to take its chances in the nest with the young of its foster parent. But not so the young of this bird. About the third day, and while it is still blind and without the vestige of a feather, the young cuckoo begins to sidle up to one of its fellow-occupants of the nest and to lift it on its peculiarly shaped flat back. The effort continues for long, and there can be no doubt from the beginning about its deliberate intent. Climbing backward up the side of the nest and using its long naked wing processes as props, it gradually lifts and gradually pushes its fellow-nestling over the side to its certain doom. The blind little imp even feels round to make sure that the deed is properly accomplished. Then it descends again, and after a rest proceeds similarly with the other nestlings. If the cuckoo is hatched out first it treats the eggs in the nest in the same manner. For hours, and even for days, the struggle continues, the end being always the same—that the cuckoo remains the sole occupant of the nest. The concentrated intent of so immense and prolonged an effort on the part of a creature so young and apparently so mindless at this stage is one of the most uncanny sights in the whole range of nature to watch.

But this is only the beginning. In a few days, and long before it is fully fledged, the bird completely fills and overflows the small nest. It now exercises a kind of fascination on the foster parents, a fascination which extends even to human beings. For when brought up by hand have you not seen the members of a whole family, yourself included, continuously hunting for caterpillars to maintain it? A bird with an evil past it may be, a bird greedy beyond compare; but a bird so gentle, so absolutely trustful of its friends, so evidently well-bred, so surrounded with an indescribable air of distinction, so evidently expecting every helpless want to be attended to—who could resist it? But a parasite withal; outside human society the most distinguished and accomplished parasite in life.

The black swifts screaming loudly and hawking for insects overhead move slowly across the sky to the west. As the light wanes in the late afternoon an effect that must have some deep bearing on the migration of birds is noticeable in this landscape where the ground is always black. In the west the sun floods the sky, the golden light shining through the transparent wings of the flying insects, and through the tops of the birches where more than one song-thrush is now pouring forth its evening melody. But all the east is sombre. The shadows fall towards it, long, dark and gloomy, and with a depressing effect which is like a leaden weight on the senses and emotions. It is remarkable in the psychology of wild life as in the psychology of peoples how insufficiently we have taken account of the deep-reaching, soul-seizing effects of waning light which is falling away from us.

It is one of the facts in the migrations of birds over which naturalists have always found a difficulty that the migrants both in the eastern and western hemispheres should in their journey to the south often begin to leave their haunts before the food supply in any way fails them, and before they have any physical want known to us indicating a coming change in the conditions of life. But students of the subject have probably not fully reckoned with the deep emotional effect on all wild nature of the waning light in the declining year, and on the uncontrollable instinct to follow the sinking sun begotten in those whose habits of life it affects.

Deep down in the psychology of peoples lies the corresponding primordial instinct of the waning day. How dying men have begged to have their faces raised to the setting sun. How we have built our houses open towards the west, intolerant of the depressing shadows between us and the sinking day. How the towns and cities of European peoples have, where possible, grown as by a natural law towards the lighted west. How it has been the west-end which has become in consequence the residence of wealth and fashion. Beneath the face of history the same sub-conscious instinct has moved to results massive and enduring. How the individuals and races, emotional, adventurous, sanguine, have ever turned their backs on the east and followed the sun. In these western lands which leaned at last against the impassable ocean they could get no further, and the successive waves of the warring peoples, pent together, were heaped one upon each other till they have become what they have become in history, and we search downward now through their records as through geological strata. And how when the ocean itself was conquered at last the flood of peoples with accumulated impulse burst once more westward over the virgin continent of the New World!

Down the valley comes the little local train, shaking the bog like an earthquake. The feeding water-rail in the deep *rhine* by the side scarcely lifts her head. The sitting lapwing still covers her nest undisturbed. From over the open comes the laughing note of the green woodpecker in the little farm orchard carpeted with flowers which nestles by the hills. Further, far distant down the valley, sounds faintly the shriek of the locomotive on the main roads tearing its way to the metropolis with ocean mails from the peoples beyond the sea. In the little station the travellers bear the marks of many occupations and of residence in distant lands. The starlings fly overhead with food in their beaks close by the spot where the coiners in the days of the Roman occupation imitated in these swamps the imperial coinage and left their coin-moulds to this day. Around lie the silent hills scarred with the works of the unknown and forgotten races. And everywhere eternal nature. How old the world is! Nay, but rather how young; in what callow infancy! The morning and the evening are as yet scarcely the first day. The conscious history of the race has as yet hardly circled the earth in its spread from east to west.

III

WILD BEES

THE wilder humble bees are a much-wronged and slighted family. Their showy relatives, the hive bees, make such a noise in the world that they have appropriated most of the attention which we have to bestow upon bees in general. The naturalist and bee farmer vie with each other in enlarging on the merits of the hive bee; the one extolling her as an example of intelligence and the other as a source of profit, they have together managed to induce us to devote to her an expenditure of printers' type which ought to satisfy the hungriest ambition for fame. But the poor humble bee, like most poor relations, has been sadly overlooked. Even the naturalist is inclined to treat her as he does ordinary moths and beetles, bestowing on her little more notice than is required to classify and describe varieties; and so the world follows suit and nicknames the poor bees in disparaging reference to their supposed inferior position in insect society—nicknames them, for the name humble bee is but a corruption of the true title bumble-bee, which—like the scientific generic term *Bombus*, from βόμβος, an imitative word—has reference to the booming sound made by the insect in flying.

The bee family has an interesting history. It is in all probability descended from certain solitary wasps which, like an existing species, were in the habit of providing living animal food for the young larvæ. Like a family of solitary wasps of the present day, they probably possessed the power of stinging their animal prey so as to paralyse without killing it, laying an egg beside this living store of food, and leaving it to be the prey of the resulting larva. The first step upwards was the abandonment of this habit, the more enlightened individuals taking to feeding the young with food disgorged from their own stomachs, the perfect insect feeding on honey or pollen. Hermann Müller states that the new race at first differed only in this habit, but in course of time, filling an unoccupied place in nature, it increased enormously, and at last formed the widely ramified family of bees as we have come to know it. The steps in the development of the family have been marked by the gradual elongation of the tongue and the adaptation of the mouth parts to honey-collecting habits, the acquisition and perfecting of pollen-collecting appendages, and the development of social instincts in some species. The steps may still be traced through surviving types, and Müller has sketched the development upward through the species of *Prosopis*, which differ little from many sand-wasps, *Sphecodes*, *Halictus*, *Andrena*, until the *Bombus* or humble bee family is reached, this being the nearest ally to the hive bee, in which the series culminates. The resulting changes which have taken place elsewhere in nature *pari passu* with this change of habits in the bee family must be left to the imagination. Nearly all the endless variety of flowers with their perfumes and colours as we know them have since been evolved, as well as most of the beautiful arrangements for bee-fertilization upon which thousands of species of plants are dependent for their existence. Truly very small causes sometimes have prodigious results! How many people know or realize that much of the variety in plants, most of the colours in our gardens, many of the perfumes on our toilet tables, much of the beauty in many of our canvases, a good deal of the poetry in our language, and even a considerable development of the beauty sense in ourselves, result from that rather vulgar historical incident dated an æon or two back, when the young of the bee family left off a taste for butchers' meat and took to vegetarianism!

Lord Avebury puts forward on behalf of the ants what he calls a fair claim to rank next to man in the scale of intelligence, if they are to be judged by their social organizations, their architectural abilities, and their relations to other animals. Their relations the bees are, however, scarcely less interesting, and most of their wonderful habits must have been developed since they acquired their social instincts, since to the acquirement and development of these most of their importance is due. For this reason the present condition and habits of the humble bees are of special interest, for here we have, as it were, the starting point where we may see the community just in process of development, the social ties which hold it together being as yet of the loosest.

Every one must, I presume, admit a general acquaintance with the humble bee. From the first sunny day in March to the last in September she is with us, industrious, contented, and entirely devoted to her own affairs, a worthy example to her betters. About forty species variously habilitated in artistic patterns in black, red, brown, and yellow are known in the British Isles, and are familiar under various names—bumble bees, humble bees, dumble dors, wild bees, foggie bees, field bees, boom bees, hummel bees, etc. Most persons must also admit a general acquaintance with her habits, and will have seen her in early spring, sleek after her long hibernation, and big with the projects of maternity, curiously spying into dark corners in search of suitable quarters where she soon hopes to be mistress of an extensive *ménage*. The desirable building site for which she is prospecting on these occasions is a retreat in the sunny side of a moss bank, or a nook in a stone heap, or an eligible rat-hole, according to the family notions on the subject prevalent among the species to which she belongs. If there are any who have not made the acquaintance of the humble bee at this stage of her career, there will be few at all events who can plead ignorance of the presence of the numerous family which she brings into the world later on in the year, the individuals of which diligently ransack most of our wild flowers, unconsciously the while providing for next year's crop of those glories of the field to delight the eye of the beauty-seeker and vex the soul of the farmer. The humble bee is, above all her sisters, nature's chosen high priestess to our indigenous flowers, good and bad alike; by her aid even the latest thistle blossom is secured against the risk of scattering its downy filaments innocuously on the wind. The humble bees, like the hive species, do not, however, thrive in all localities; this is perhaps due to the absence of particular flowers in certain districts; it is remarked that the hive bees will not prosper in a locality which does not produce an abundance of white clover—a plant upon which they largely depend for supplies during the swarming season.

No one is likely to confuse the humble bee with the hive bee, the smoother body and small size of the latter being in strong contrast with the large rough-looking and very hairy body of the former. The thick covering of hair on the body of the humble bee, besides its use in pollen collecting, is intended to be a protection against the weather, and it indicates the climatic distribution of the family. The genus *Bombus* is essentially a cold climate form, and it is particularly associated with the north temperate zone. The bees have, however, managed to extend themselves far beyond these limits; their familiar boom being heard all round the world, and at both sides of the equator. In Europe they fly as far north as Lapland, and in Asia they extend from Northern Siberia to the confines of India. Most of our species are found in North America, and some of them in the southern continent. They are; however, entirely unknown in Australasia. The native bee which takes their place in Australia is a puny stingless weakling, resembling the aborigines, even to the extent that it is being rapidly exterminated by a higher civilization, progress being represented by the hive bee imported from Europe. But to return to the history of the humble bee. When the queen issues forth in spring, after her long winter sleep, to found a new community, she, as soon as she has fixed upon a suitable spot to build in, collects a small store of honey and pollen as a provision for the first members of the family which she is about to bring into the world. When the queen bee of the hive species has made up her mind to start on her own account she issues forth from the parent hive followed by thousands of ready subjects. The swarm is, indeed, already a miniature state with a perfectly organized government, and with suitable quarters the machinery of government works smoothly from the first. With the queen mother of the humble bees the case is entirely different. As yet there is no state, and she has no subjects; she represents the whole in her own royal person; " *'Létat c'est moi!* " is just now in her case literally as well as figuratively true. She lays at first a few eggs, and continues depositing at short intervals from half a dozen to a dozen at a time. These eggs all produce neuter workers, and directly the first bees hatch out they relieve the queen of all the inferior duties thrust upon her by the necessities of the situation. They are soon in the fields, collecting supplies for the increasing wants of the young colony, and the queen having resumed her proper position, she, like her sister of the hive species, in future devotes her attention solely to the production of eggs, her subjects taking charge of all the duties of nursing, food-producing, house-building, and defence.

My observation of the colonies which I have kept has always dated from about this stage. I have never been able to induce a queen to commence housekeeping under observation. I have kept the young queens from my nests till the following spring, as well as those captured in the fields late in the autumn, and I have also tried with two caught in the early spring, but in all cases I have failed. My colonies have always been taken from the meadow.

On the sill of one of the windows of the room in which I am writing is a shallow wooden box open at the top, in which is established a flourishing colony of humble bees, one of four which I have kept since last May. This is the third season in which I have had colonies of the humble bee under observation, and I have found them to thrive unexpectedly well in London. This colony in particular has flourished. Beneath the round moss-covered dome in the centre of the box it has stored up a supply of honey and pollen large for its kind, and, having brought into the world a numerous progeny of workers during the season, it is now hatching out an unusually large number of young queens and males to carry on the species next year. Bohemians as these bees are, it is curious and not a little interesting to see the intelligent active little creatures so much at home here, and to watch them coming to my window over the houses and trees laden with spoil. It is always a pleasing duty to bring deserving claims to notice, and I shall feel that I have discharged a duty if a pretty intimate acquaintance of the family affairs of the colonies which I have kept under observation will enable me to say anything which might tend to the removal of the badge of inferiority which by common consent seems to have been bestowed on the humble bee.

Those colonies which I have this year are just now (July) in the heyday of their prosperity. The one to which I have already referred is arranged for observation. The bees resent any interference with the nest during the daytime, but here beneath the shade of my reading-lamp I can remove the mossy dome, lined inside with wax and waterproof, exposing the busy scene underneath, without any hostile demonstration being attempted against me. After a few moments' perturbation the bees do not seem to miss the covering, and the work of the nest goes on as usual. It is an interesting sight. The bees are all employed in some way, some excavating round the edges of the nest to make room for the growing bulk within, others attending to the multitudinous duties connected with the shaping of cells and cocoons and the wants of the young larvæ. Even those which appear to be lazily stretched at full length over the cells are not lying luxuriously at their ease as might be thought, but are helping to keep warm the young larvæ within, assisting their development by a kind of incubation. The so-called honeycomb in the nests of the humble bee is a poor affair compared with the beautiful structure manufactured by the hive bees; it is not really of the same nature, but consists simply of the cocoons of the young insects irregularly fastened together. Some of those in the nest still contain the young brood, but others out of which the young insects have emerged have been used for the storage of honey. Those dark brown protuberances affixed in some places to the cocoons, looking like small accumulations of wax kept in reserve, have an interesting history; they contain the young bees in various stages of development before they spin their cocoons. The queen generally lays in one of the little recesses formed where two or more cocoons join, and the eggs are afterwards covered with a thin layer of wax and pollen, which the bees add to as the larvæ inside increase in bulk. Instead of each grub occupying a separate cell, like those of the hive species, many are here wrapped together under a common blanket. It is easy to expose the eggs or larvæ underneath by raising the thin covering of wax with the point of a long needle, and it is amusing to see how the bees crowd excitedly round, and in a few moments repair the damage under one's eyes. They show some suspicion of outside influence, and even bite inquiringly at the needle as if not altogether satisfied that its innocent-looking exterior may not be a cover for hostile intentions. Some time ago, being curious to see, amongst other things, whether the young of the hive bee could be reared under the same conditions as those of the humble bee, and *vice versâ,* I made some trials with the eggs and larvæ of each. Among other experiments I placed some larvæ of the hive bee in one of the nests of the humble bees. Having carefully removed part of the waxen covering of one of the little groups of larvæ, I placed a grub taken fresh from the hive amongst the others, and covered the whole roughly up again, expecting that as usual the bees would complete the repairs, and so seal up the intruder with the others. I was, however, disappointed; they were not to be cheated in this way, and they would not repair the broken wax until they had smelt out the stranger, whom they dragged out and carried outside the nest, after which they repaired the breach in the usual way. Several times I tried, but with no better success. Unlike the ants, who will rear the young of other species, these bees would not allow the strange children of their betters to be fathered on them, and the helpless little aristocrats were always detected and dragged ignominiously out. As some species of the humble bee tolerate a kind of cuckoo bee (*apathus*), which lays its eggs in their nests to be hatched out with the offspring of the legitimate proprietors, whom they much resemble in appearance, I was not altogether prepared for this intelligent opposition to my ideas. Thinking that I might have better success with the eggs, I took some fresh from the hive and placed them amongst a little group just deposited by the humble bee queen. The bees at first appeared to be rather puzzled at these eggs. One or two of them took them up somewhat aimlessly, and again replaced them as if they hardly liked to openly accuse their sovereign of misconduct, which they seemed to suspect. After a little hesitation a decision was arrived at. *Natura non facit saltum* was surely as safe a motto for bees as it is for philosophers, but instead of carrying the eggs out of the nest as they did the larvæ, the bees, one after the other, proceeded, apparently with considerable relish to nibble them—a relapse into barbarism which, after ages of æsthetic culture, was quite startling—and so appreciative did they become of the flavour of these new-laid eggs that they would soon accept them readily when I offered them at the end of a needle. Once or twice I think I succeeded in smuggling some of the strange eggs in with the others, but, whether it was that the bees afterwards detected them or that they were hatched out and the strange surroundings and low company proved too much for the refined tastes of the hive grubs, they never came to anything.

The wax which the humble bees use to stick their cocoons together, to plaster the inside of their nests, and sometimes to form rough cells to hold honey, is very dark, almost black when compared with the beautiful white material secreted by the hive bees. I think they mix earth with it, for I soon found that my colonies did not flourish so well unless I provided a supply of earth kept constantly moist within reach of the bees, and in this they were constantly burrowing.

The humble bees never swarm ; there is nothing amongst them analogous to what happens when a colony of hive bees with the queen at their head issues forth from the parent stock to found a new community. About the middle of July a colony is at its best. Up to this time nothing but workers have been produced in the nest, and the bees will have garnered a supply of honey and pollen which, in the case of the underground species, where the colonies are larger, will have assumed considerable dimensions. This is all, however, but a means to a great end, none of the present occupants of the nest having the slightest interest in its prosperity. They have borne the heat and burthen of the day only that others may reap the fruits of their labours. About the beginning of July a change comes over the queen. Hitherto she has produced only neuter worker bees or imperfect females, but now, whether by instinct or necessity is not yet clear, she commences to lay eggs which produce only males and perfect females or queens in about equal proportion. When these attain to maturity the dissolution of the commonwealth is at hand. Neither the males nor the young queens take any interest in the affairs of the colony, and they both forsake the nest soon after they are able to fly. The males pass a brief roving existence, exhibiting a marked spirit of masculine independence, stopping out all day and all night too. They spend the sunny days in looking for their mates, till having fulfilled their mission in the world, they, like the workers, all die at the approach of winter. Only the queens remain, and these wander about till the cold weather warns them that it is time to search for a warm corner, where haply they may survive the winter to start the following year on their own account.

It is generally supposed that the old queen does not survive a second winter. The queen bee of the hive species it is well known lives and remains fruitful for several years. I have often missed the old queen from my nests towards the end of the year, and have often wondered what had become of her. One morning early in July last year I was watching one of my colonies at work when I was surprised to see the old queen come out unattended, and after a little hesitation fly away. I watched for a long time but did not see her come back, and as far as I could learn she did not return during the day. Next day I opened the nest. The colony was in a very flourishing condition; some young queens had just been hatched out, and there was a good deal of young brood in various stages of development, and some eggs which had not been long laid, but,. as I expected, the old queen was missing. She never returned, but the affairs of the colony went on as I should have otherwise expected until all the bees dispersed at the usual time. It is likely that the old queen is one of the first to leave the nest towards the end of the season, and it is not improbable that after spending the autumn like the younger queens she should in some cases survive till the second season.

As far as my observation of the humble bee goes, I have found the individuals more intelligent than those of the hive bees. This may seem strange considering the work and the wonderful social organization of the latter. Yet it is doubtless in result, to quote from Mr. Herbert Spencer, a question of altruism *versus* egoism. The specialized instincts of the hive bee have been for countless generations developed on the strictest lines of altruism, that is for the benefit of the colony rather than of the individual bee. A glaring example of this altruism carried to its extreme limits is witnessed when the hive bee, Spartan-like in its public spirit, but pathetic in its stupidity, sacrifices itself on the smallest provocation for the good of the commonwealth, when it inflicts a slightly more serious wound by leaving its barbed sting, which it cannot withdraw, rankling in the flesh of the intruder, and dying itself from the injury caused by the loss of it. So it is with most of its instincts; they have been developed and specialized for the good of the community and do not necessarily imply what might have been looked for as a corresponding degree of intelligence in the individual. When the humble bees long ago branched off from the parent stock, the individuals of the species were doubtless still so sunk in barbarism as to be quite incapable of even unconsciously entertaining any designs which had not a direct bearing on the interest of number one. The habits of the humble bee, still living a single-handed existence for a great part of the year, have made it necessary for her to retain a good deal of this primitive wisdom, and even to cultivate it. The development of the two branches of the family has been on entirely different lines. The humble bees in fact have not yet learnt to sacrifice the individual to the community, and despite their very rude social economy, and the popular prejudice against them in consequence, it is not at all improbable that we may have yet to allow that the individual humble bee is in advance of her cousin the hive bee in "general intelligence."

I need not refer amongst other things to the intelligence displayed particularly by the members of the underground species, in their ingenious plan of getting at the honey in some flowers by piercing the corolla, a habit which the hive bees are ready enough to take advantage of without having the intelligence to imitate it. There is a sense of individuality about the humble bee which it is hardly possible to attach to a single bee of the hive species. One sunny day in March I captured a large female of the species *Bombus terrestris* on the willows in the wood above Weston-super-Mare. Taking her to London with me, I placed her in an empty nest in which I had kept a colony the previous year, and having filled part of the empty comb with honey and given her a supply of pollen I was in hopes that she might be induced to rear a young family under observation. I was, however, disappointed to find that beyond helping herself to the honey she evinced no desire to take up her quarters in the vacant nest, and eventually despairing of success in my endeavour to reconcile her to her new surroundings, I gave up the attempt and let her have her own way. She then spent several days in beating against the glass of my windows in the endeavour to get outside. Being much from home at the time I took little further notice of her then, and soon missing her, I concluded she had found her way through the open window. One warm day some weeks after, as I was quietly reading in the room, I was a little surprised, on hearing a slight scratching noise near me, to observe her, engaged upon her toilette, perched beside me on the table-cover, out of the folds of which she had evidently just emerged. She had apparently take up her quarters permanently there, for after a few short flights round the room she returned and eventually retired to her old quarters for the day. This was the beginning of a long and interesting acquaintance with my humble friend. Nearly every day during the early summer she came out, and her behaviour on these occasions was very curious. Her early experience with the windows had evidently made a great impression on her, and she never attempted to escape that way now. Sometimes, indeed, after a long interval, she would fly towards the light, but before she reached the glass she invariably turned back from what she had evidently come to regard from painful experience as a delusion and a snare. She made short trips about the room all day, generally retiring in the evening to the folds of the table-cover. On these occasions her interest in the objects in the room was of a peculiar kind; the long confinement must have acted upon her as it often does on prisoners, when it leads them to take an unusual interest in objects they would not otherwise have noticed. All the bright objects seemed to attract her. The brass handles of some cupboards greatly interested her, and the polished knobs of a set of drawers were also a source of great attention. From time to time she would alight on one of these latter, and having walked all over it would fly to another and another without again alighting, as if she had been visiting flowers. She was on the most friendly terms with me, but I felt rather slighted to find that her interest in me was principally centred in the buttons of my waistcoat, which were made of some polished material which doubtless attracted her like the knobs of the drawers. A great object of interest to her also was a bookcase, the backs of the books in which in various coloured bindings, labelled in gilt letters, were an unfailing source of interest. Here again she rarely sought to approach the glass, but remained on the wing outside while she studied the gilt titles beyond. It seemed to me strange that she should return again and again to such a profitless occupation; yet she made many visits in the day to the same objects. I am inclined to think that in her visits to the bookcase the books which received most attention were those in green covers (there were very few in blue), although the rather florid lettering of some of the poets in red binding also seemed very attractive. The greatest object of attraction in the room was, however, the keyhole of the door. Into this she would try to squeeze herself, and failing, would alight near it and walk round and round it. It is no impropriety to say that the instinct which moved her here was essentially a feminine one, for she doubtless associated the small dark opening with the entrance which the females of her species usually choose for their underground nests. Acting on the hint, indeed, I afterwards took the discarded nest before mentioned, and placing it in a small wooden box entirely covered over, in the side of which I had previously cut a small opening, I fixed it in a prominent position and soon had the satisfaction of finding her enter and take permanent possession. She would not, however, be induced to breed, and one morning towards the end of May I found her dead in the nest, although she had a plentiful supply of food within reach.

During the few months this bee was with me her general relationship both to her surroundings and myself evidenced an individuality which we are not accustomed to associate with the members of the insect family. She certainly regarded me without fear, and I am inclined to think with some degree of favour. I used occasionally to stroke her with the end of a light feather, and she used at times to show her keen enjoyment of this by stretching at her full length during the operation, often putting herself in this position before the feather had touched her. At other times, very curiously, she would not submit to be stroked at all, and seeing me make preparation to do so, and while yet the feather was some inches distant from her, she would throw herself on her back and scratch and bite viciously, although she would allow me to take her in my hand without attempting to sting.

As already mentioned, one of my colonies contains an unusually large stock of young queens this year. I have fed the nest liberally during the year with pollen taken from the hive bees, and this may account for the superabundance of royalty. Royalty amongst the bees is not a matter of birth, but of breeding, or to speak more correctly, of feeding. When the hive bees want to manufacture a queen to order they take an egg or young larva, which under ordinary circumstances would develop into a neuter worker, and by special feeding and the necessary enlargement of its cell, it blossoms forth into a fully developed queen. In the bee-hive, indeed, the plebeian worker may at any moment in her youth have greatness thrust upon her, for as in the great Republic, the meanest citizen (if caught young and of the feminine gender) is eligible for the highest honour which the state has to bestow.

With the humble bees royalty is doubtless manufactured in a similar way. Any one who has disturbed a nest towards the end of the year may have noticed that the workers are sometimes of various sizes. Those produced at the beginning of the year are often only slightly smaller than the queens, but towards the end of the season I have seen worker bees little bigger than house flies produced in the same nest with those large workers. It has been stated that the difference in size in these cases is due simply to difference in feeding during the larva stage, and if this be true, it means that those small workers produced later in the year are the stunted victims of the process of gorging to which the crop of young queens is subjected, the workers being robbed of their proper quota of food to supply the wants of what may not inaptly be called a bloated aristocracy.

This case of neuters or sterile females among the social *Hymenoptera* is one of the deepest interest in all its bearings. Darwin, while explaining in the "Origin of Species" the action of natural selection here, has left it on record that the case presented to him difficulties which at first appeared insuperable, and actually fatal to the whole theory of natural selection. The question of sex among the humble bees is most interesting, as tending to throw some light upon the subject where it presents more difficulties, namely, amongst the hive bees. Amongst the humble bees the differentiation of sex has hardly begun. The queen performs the duties of an ordinary worker for part of the year, and the worker female differs little from her in anything beyond what may easily be understood as under-development consequent upon less generous feeding during the larva stage. But with the hive bees the divergence is far wider and more significant, involving not only difference in development but in instinct, and what is more important, in structure. The queen of the humble bees, like the neuters, possesses pollen-collecting appendages, and a curved almost unbarbed sting (which does not remain in the wound), which heritage she of course still transmits intact to her royal descendants and to the neuters. Now the queen of the hive species, besides differing altogether from the neuters in instinct, has lost, with other slight peculiarities, the pollen-collecting appendages, but she still preserves the power of transmitting them to the neuters, while on the other hand she has retained her curved unbarbed sting, but strange to say has acquired the power of transmitting an improved and more deadly weapon to the neuters. In the case of the neuters of the hive species it is interesting and not inexplicable that the peculiarities of instinct and structure which are correlated with sterility should be developed in them by the principle of natural selection acting on the community, though transmitted to them by the queen in whom such peculiarities have never been developed. This may be explained; for as in the course of time modifications of structure and instinct in the neuters were found to be advantageous to the community, there was a tendency for the fertile females in the communities in which those modifications were most pronounced to flourish, and so transmit to their fertile offspring a tendency to produce sterile members with the same peculiarities. Yet the most wonderful feature of the case remains to be mentioned, namely, that in the hive bees those peculiarities which the fertile female or queen transmits to her offspring can be controlled and profoundly modified simply at the will of the worker bees by the course of treatment to which the young insect is subjected while in the larva stage, so that from the same egg may be produced either an ordinary neuter—with pollen baskets and barbed sting—or a queen without the pollen-collecting appendages, which would be useless, and without the barbed sting, which would be a dangerous if not a fatal equipment.

That natural selection has been the *causa efficiens* in bringing about this remarkable combination of peculiar endowments working together for the good of the species, there can be no reasonable doubt; nevertheless, the mind cannot withhold its tribute of admiration when we consider the exquisite adjustment of means to that end here presented, and reflect what the steps must have been before the present advanced stage towards perfection had been attained.

The males of the humble bee are an interesting section of the community. They differ in colour from the females and are more brightly marked, but it is peculiar that there is no permanence of pattern, the males of the same species differing in the arrangement and intensity of the colours displayed. Besides other structural peculiarities they possess much longer antennæ than the females, and like the drones of the hive bees they are not armed with a sting. Comparing them with these latter one cannot help being struck, here as elsewhere, with the greatness of the penalty which the individuals amongst the hive bees have had to pay for the social organization which has contributed so much to the success of the species in the great struggle for existence. The male bee of the *Bombus* family is still far from having become the helpless pitiful creature which we find his male relative the hive drone to be. True, nature has already made it clear that he is a creature of infinitely less importance to her than the females who are to carry on the species; but beyond this he is not to be despised. Although he has no sting he submits to no dictation from the neuters in the nest, for he leaves it immediately, and what is more important, he is under no necessity of returning, for he can forage for himself among the flowers, and he is not therefore like the hive drone, reduced to that most dismal of all necessities—sponging on unwilling relations. He enjoys himself in an independent way while he can, and he is not likely to remain long unmated. The poor degraded hive drone suffers much in comparison; he has fared badly while the great principles of free competition and *laissez-faire* were winning for his species a worthy place beside the ants. The hive drone is produced and maintained by the colony for one purpose, and all his instincts which do not tend directly or indirectly to further that single purpose have been retrograding. He has lost most of what intelligence his kind ever possessed; he has lost the power of seeking his own food, being helplessly dependent on the earnings of the colony. Of his miserable life the bees are so contemptuously generous that for one queen which requires to be mated they generally produce at least one hundred drones, ninety-nine of which, excluding accidents, live to be ruthlessly slaughtered by the bees at the end of the season. The favoured one which meets the queen on her marriage flight pays the penalty of being chosen to such a lofty destiny. The queen returns to the hive alone, and during her life remains true to her first love, but her elect returns no more; he has been the hero of a love tragedy, and even *in articulo mortis* has become the father of a mighty host. This peculiar fate, which happily is rare in the animal kingdom, is, however, said to be shared by the male of the humble bee, but here I cannot speak from personal observation.

Has the present condition of the humble bees and their evolution to a higher social development any bearing on other questions? The family or clan stage of social development is represented in the village communities of the humble bee, still held together by the loosest ties. The independence and welfare of the individual is still preserved, and the community still largely exists for the individual, and not the individual for the community. But with the hive bees the individual has ceased to be of much account ; even its life is wantonly expended, as, for instance, in order that the colony may secure the infinitesimal advantage derived from the slightly more serious wound inflicted when the worker leaves her barbed sting in the flesh of an enemy ; one-half the community (the neuters) are unsexed, and the other half (the drones), while preserving their sex, have lost nearly everything else to become the degraded victims of the meanest kind of slavery. But the species has prospered, the government is highly centralized, and the state is rich, populous, and powerful beyond comparison with its less civilized competitors. What are the lessons ? Has progress been dearly bought as we should count the cost ?

IV

EELS

NOVEMBER is the season when there is silently taking place through all the broad waters of the land a pilgrimage which exceeds even the annual migration of birds in the interest which attaches to it. The eels are seeking their spawning grounds. It is only within the last few years that science, in the case of the common eel, has found the clue to one of the problems of natural history which for long resisted all attempts to explain it. Up to the date when, some years ago, a paper on the subject was read before the Royal Society, the life-history of the eel remained a subject of mystery and uncertainty. Now that the facts have been in some measure pieced together by patient investigation, the reality has outstripped the imagination of the naturalist, and the life-story of our familiar eel, soberly recounted, reads like a page from the "Arabian Nights." Last spring, as the present writer sat swinging his legs on a low bridge over a river in Somersetshire, there were to be seen in the water beneath thousands of little eels wriggling up-stream in constant procession. It was a sight which was to be witnessed at the same time in many other rivers. The eels in the spring ascend in this way against the current in the streams all over the country, wriggling through grass and weeds and even climbing damp walls under the instinct which drives them. Every river and lake throughout the country, and even the smallest isolated pond has its eels, and the question has always been: Where do they come from and whither do they go? All kinds of stories are current amongst country people as to the origin of eels from other forms of life, or by spontaneous generation. The great mystery, however, twenty-five years ago in scientific circles was: How do eels produce their young, and where do they spawn? Eels, it was well known, remained years in the same waters; they attained a large size; they had even been kept under observation for twenty years or more. But no one had ever seen an eel containing spawn or producing young.

It appears now that all the large eels return to the sea to spawn after they spend some years in fresh water. This, however, is not the most remarkable fact of the case. In noteworthy researches made by Grassi and others in the Mediterranean, it was discovered that the spawning of our common eel takes place at considerable depths in the sea, probably never less than two hundred fathoms. The eggs of the trout and salmon will only hatch out in the shallowest water. But pressure at this great depth appears to be necessary to the vital functions in the production of the young eels. It is commonly known in this country that the big eels go down the rivers in the autumn. Those that are caught on the journey are usually observed to be undergoing a curious change. Their eyes have become larger and their bodies covered with silvery scales. The eyes are now known to reach an enormous size when the eels reach very deep water. A glance at the map of the sea surrounding the coasts of Northern Europe will show what an extraordinary interest has begun to attach to the life-history of the eel that so familiarly takes the worm with which the schoolboy baits his hook in every little stream and inland pond. There is, for instance, no depth of two hundred fathoms in the North Sea or in the English Channel or anywhere near our coasts. All the eels that come down the rivers of the British Islands and of North-Western Europe appear to be making for a region farther out in the Atlantic. A considerable distance west of the British Islands young eels have been found in water of great depth, and it is apparently from such a region that the young eels return which ascend our rivers in spring. Eels on their migration go down the rivers in the autumn generally with floods. They are caught for the markets in vast numbers at such times. They generally move at night and they seem to prefer stormy weather. As in the case of the migration of some birds the males precede the females in this journey towards the depths of the ocean. The usual stay of the eel in fresh water seems to be about five or six years. We never see the mature form, for the eel which we know develops into a creature of very different appearance when it reaches the rendezvous in the deep waters of the ocean where it spawns, and to which it is drawn by these strange forces of life.

It is curious how all this seems to fit in with much that was formerly known. Seventy years ago a paper was read before one of the Natural History societies in Edinburgh giving accounts of the habits of some tame eels that were kept in a pond. It was said that in the autumn they invariably became very restless, and took every opportunity of the pond overflowing from rain to get out. It was noted as a fact of interest that the eels on these occasions were without exception found travelling over the surface in an eastern direction, that being the direction in which the sea lay. The observation was the more interesting as it was unaccompanied by any theories as to why the eels should want to get to the sea. They were indeed supposed to breed in the pond, and the fact that young eels had been found there was given as proof of the truth of this supposition.

As the problem of the eel has been explained we would appear to have in its life history the reverse of that of the salmon. The interest of the eel is, however, far greater. There are many points also at which the parallelism entirely fails. It is a curious feature, for instance, that the mature eel never seems to return to fresh water after it revisits the sea. All information at present seems to point to the conclusion that the eels die after spawning in the depths of the ocean. Unlike the salmon, the eel never seems to make more than once in a lifetime the journey to meet its mate. Another interesting point is that, although the span of existence of the eel which thus completes its life-cycle appears to be comparatively short, it may apparently be indefinitely prolonged in certain circumstances. Eels that continue in fresh water remain barren. But they will live indefinitely. They have been kept in confinement in ponds thirty and forty years. They obtain a great size in such circumstances, feeding voraciously in the summer, and, as is the habit of eels, burying themselves in the bottom and remaining in a more or less torpid state during the winter.

The mystery of the migration of eels remains. It has indeed only been added to and deepened by these observations. The young elvers which ascend the rivers in countless millions in the spring-time are already a year old. They have, indeed, by the time they reach fresh water, travelled far through the trackless waters of the oceans. How do they find the way? In the case of the migration of birds there is always the suggestion that the young birds have found the way by accompanying others that have made the journey before. But here this explanation fails us. The parent eels never return. The little eels, which are about two or three inches in length, and which have already undergone metamorphosis in developing from the egg in the deep sea, can obviously possess nothing in the shape of memory to direct them. Yet they press onwards to their destination with an unerring sense. This one may readily observe if it be sought to interrupt them or to place any obstacles in their way. The common eel is found over the greater part of Europe, and if the development of the eggs, as appears to be proved, will only take place at considerable depths in the ocean, it is evident that journeys of immense length have to be made by the parent eels in returning to their breeding-places, and then by the little elvers in seeking the rivers and inland waters. It seems one of the most remarkable of the facts of migration that we should have thus to think of the eels of the countries bordering on the North Sea, after their term of years in fresh water, finding their way out of the ponds, down the rivers, and along the dim vistas of the sea to meet their mates in the still depths of the distant Atlantic. The large eels that go in the autumn have been traced to some extent in the course which they take. And we see every spring the little eels which return from the depths, after they have been already a year in the sea. But what a sight it would be if we could see it, this meeting-place of the hosts of eels from many countries which have come to spawn in the depths of the ocean.

V

HARES

THE hare in this country usually begins to breed in March, and by this season (August) the members of the first litters are well advanced in growth. The young are placed amongst the tufts of short cover in open grass land, and even after they are grown to a considerable size they nearly always try to secure safety by concealment rather than by flight. The young hares in their form in the grass constitute one of the most characteristic sights in nature, although it is one much less generally observed than might be supposed. The leverets are usually to be found in pairs, and they nearly always lie head to tail, and rarely with the heads together, probably an instinct inherited for purposes of better concealment. When they are in this position concealed in the grass, one may walk round and round them, and do everything but sit on them or tread on them, while the timid little creatures will never move a muscle. The present writer, when photographing, has bent down the grass at the side of a well-grown pair, so as to catch the reflection of the light on the eyes, almost brushing the fur of each in the act, while they still remained motionless.

There is no more remarkable type in nature than that of the hare when it is considered in relation to its specialization for speed. The wonderful symmetry of the greyhound, one of the oldest of the dog types bred by man, is but the corollary, through artificial selection, of what had been attained in the case of the hare by longer ages of natural selection. Our common wild hare, seen in an attitude of attention, is a beautiful creature, displaying in every movement the nature of the history which has produced it. The large, bright, intelligent eye, so different from that of the rabbit; the deep, cup-shaped ears, capable of being bent in any direction to form a receptacle to catch the slightest sound; the well-correlated movements, showing intention and intelligence at every turn; the body itself, with its marvellous blend of protective colours and its suggestion of speed in every line; the long and remarkably built hind legs, moved by the powerful muscles above and tapering to the slender feet; the characteristic leaps and gambols of the creature, which is capable of tucking or folding itself when at ease into a space only one-fifth of its fully extended length; and, lastly, the pervading consciousness, manifest in all its actions, if it be in the least suspicious of being watched or pursued, of the betraying scent given off by its body, all form a blend of qualities irresistibly suggestive of the untold ages of stress and selection out of which the hare has come. The principal natural enemies of the hare in this country are the fox, the weasel, and the polecat. The behaviour of the hare on being hunted by the weasel or polecat much resembles that of the rabbit. It is in both cases inexplicable. The hare, which could easily far outpace its enemy, shambles uneasily forward; it allows its pursuer to overtake it, and it often, like the rabbit, squeals as if in mortal terror long before the bloodsucker fastens upon it. The hare breeds in this country several times in the year, and with all the protection which is afforded it, the destruction which goes on must be immense or otherwise the land would be overrun with hares. The mortality amongst the early litters is great, particularly in cold and inclement seasons. It is indeed a singular fact in natural history that the hare, every detail of whose body is related to the fact that it numbers nearly every beast or bird of prey amongst its enemies, should be so delicate in constitution. A cold night in spring, as many an observer must have noticed, kills numbers of the young in the early litters. Shock or slight injury is also readily fatal, even when growth is well advanced.

The stories told about the intelligence of hares when being hunted by dogs are innumerable. The animal will return over its scent, cross and recross it with springs, and make off at right angles. It will go down one side of a hedge and then up the other, passing its pursuers with only the screen between. It will take to water or endeavour to lose the scent amongst domestic animals. We have even known it to jump on the shoulders of a man when hard pressed. Even the little leverets, as they hide in the grass in the spring, seem to have a highly-developed sense of the necessity for cunningly meeting the dangers to which they are exposed.

If a pair are marked in a form the entire lack of motion only lasts while they are under observation. It is preliminary to immediately shifting their quarters as soon as they feel safe to move. They will invariably be found to have disappeared if one returns to look for them soon after. Our common wild hare becomes a delightful companion when tamed. One which the present writer obtained at an early age grew as familiar and playful as a kitten. It would sleep in a basket by the fire during part of the day, and come out for its gambols about the room as evening approached. The animal, during some months it was under observation, displayed in all its movements how deeply the hunted life of its kind had left impress on all its qualities, and how watchfulness as to the necessities arising therefrom formed as it were the dominant character of the hare's mind. If it was introduced to a strange room, for instance, its first behaviour was always the same. Although it showed no fear if accompanied by any one it knew, it invariably continued in a state of mental tension until it had satisfied itself on one point. It made sure that there was a place to which it could retire at will, and if possible remain screened from view. Then and then only would it relax into playfulness and ease. One of its continual frolics was to come out of this hiding-place and then pretend to be chased back at full speed. On some occasions the mixture of shyness and daring was most fascinating to watch. There was apparently a mental process going on in which native timidity was being continually corrected by the assurance that everything was quite safe. It was boldest and most familiar at night. It would then come on the table while writing was going on and stretch itself out at full length, or sit and watch with four legs tucked underneath like a cat.

It is a moot question whether the hare is a rabbit which has taken to the open or the rabbit is a degenerate hare which has obtained comparative safety by taking to a stupid life in the earth. It is an interesting fact in this connection, and one not often remarked on by observers, that a hare, if it finds an obstacle it wishes to get rid of, will naturally scratch with its front legs with considerable strength and with exactly the same movement as a rabbit. Thus, although the hare lives in the open grass country, never takes to earth, and much dislikes ground infested by rabbits, it has to all appearance latent in its muscles the beginnings of an instinct which might be developed into the rabbit's capacity for burrowing.

VI

A MIDSUMMER NIGHT

IT is one a.m. We are on the open chalk downs under the stars, and twenty miles due south from London as the crow flies. The low summer moon, which has been but a few hours above the horizon, is already sinking away in the south-west. There is but little light, for the pale yellow beams do not illuminate ; now, even before the dawn has come, they are waning, and a ghostly air has settled upon the almost invisible landscape. The northerly breeze has come through the wood which meets the sky in the foreground, and the aroma of leaves, still in all their delicate summer freshness, lingers on the night air. The distant bay of the watch-dog comes over the hills, to be answered by another still farther away, and yet now by another in the valley below. But the sounds themselves are part of the solitude ; they seem only to increase the silence.

Under the clear sky the heavy dew has made the grass dripping wet, and in the uncertain light it is difficult to keep to the steep pathway through the upland meadows. In the low ground below, where the trees rise spectre-like through the mist, the railway runs. It is but a few hours since the roar and crash of wheels echoed up here, and the tail-lights of the Continental express flashed through the trees ; but shadowy and unreal seems the world to which such life belongs, a part of a far-off existence which has no touch or communication with these rural fastnesses. It is a silent land. Celt and Roman and Saxon alike have carried highways of the world through it. But it is still silent ; now, as ever, the life of the highways tarries not in these solitudes which sleep between London and the southern sea.

Chur-r-r-r !—distinct and eerie, the sound comes up the hillside, the air vibrating with the harsh rolling note. Now it is answered by a similar sound, and the belt of small oaks and bracken below seems suddenly possessed by a troop of invisible spirits. It is the fern-owl, or night-jar, calling to his mate—a sound which has caused a growth of superstition to follow the bird into every land in which it has travelled. The female, who nests on the ground, is usually sitting when the male makes the night air thrill with his strange note. The bird is heard here only about this season. Out of the unknown it comes with the rising year, and thither it returns with its decline, reaching here on the crest of that great migratory wave of life from the south, of which we know so little, and which now, almost with the summer solstice, will turn again as mysteriously as it came.

Slowly the splendid summer night opens out as the ground still rises. Far away in the north, in the direction of London, a soft opal light hangs upon the horizon. It is the fringe of twilight from the midnight sun circling below the horizon, though it is still more than two hours to sunrise. The moon has almost ceased to shine, but the planets burn more brightly as the light wanes, and a deeper hush seems to fall upon the darkening landscape. Hark ! in the still night air at this altitude the ear catches now for the first time a solemn undertone of the night. It is like the subdued echo of the surf, but from a shore so distant that the sound is here only the gentlest sigh in the air ; the ear strains after it when at times it seems to melt back again into the silence. The ground here is the watershed between two rivers, the northern Thames and the eastern Medway. It has been raining heavily during the past week ; every little rill is full, and the river in the valley below is still in flood. It is the faint sound of the plash and fall of many waters which reaches here in the stillness. This is that voice which, once heard at night on the open hills or moors, is never forgotten ; that sound which, more than any other audible to human ear, suggests the infinite—

> The sound of streams that swift or slow
> Draw down Æonian hills, and sow
> The dust of continents to be.

The pathway through the fields runs close to the hedge now. The scent of white clover comes down the breeze. In front, where the ground rises highest, the Southdown sheep lie huddled against the sky-line. They have given an historic name to a breed famous for its mutton ; yet even in such descendants survive the instincts of long-forgotten ancestors. It is the highest spot of the pasture they have chosen to rest in, and they lie with noses to the wind, waiting, they know not why, for an enemy that will never more disturb the slumber of their degenerate lives. Faint brushing sounds come through the grass; shadowy forms which the eye does not catch seem to move before; a hollow, sepulchral double knock comes from the depths of the hedge: it is only the angry, warning stamp of the rabbits that have been disturbed feeding.

As the road goes north the scene changes. These rolling chalk downs, with the deep combes nestling at intervals between, have given trouble to the ancient road-makers: now the track mounts suddenly and steeply, and in an instant descends again almost precipitously. Here the hills have closed round again, the breeze is no longer felt in the valley, and the shadows seem to come closer. The long, lush grass, almost ripe for cutting, still stands by the road, and the green wheat, already in the ear, makes a sombre gloom on the southern slopes under the hazel copses. Crake-crake, crake-crake!—far and wide the sound echoes through the still air. It is not a stone's throw off now, and it comes from the thick cover by the roadside, harsh, loud, and strident, drowning all other noises of the night. It is only the love-note of the land-rail, one of the most familiar of all the night sounds in this strange wanton honeymoon of our Northern year, when for a few short weeks all nature stirs and glows and seeks to utter herself of a life that passeth understanding. Thus still for a little does the male bird cheer the female as she sits on the eggs. Yet a few weeks more, and he will be no longer heard; for he will change and relapse into silence and other moods when the young are hatched out. The sound ceases suddenly now, only to render audible a similar note in the distance. When it is renewed, after a short interval, the bird has moved. He travels quickly through the long grass. Well do you remember how in other days you hunted him, what good sport he made, how fleetly the long legs carried the slim, brown body, how loth he was to fly, and how heavily he rose. The country-people said, indeed, that his wings were of little use ; that, left to himself, he never used them ; and even that he shed his feathers, and slept through the winter in the rabbit-burrows. Yet not the least of nature's mysteries are the now well-established wanderings of this familiar land-rail of our homestead meadows. By what strange routes has he been tracked over land and ocean with the waning year, south along the Nile valley, and even across the equator into southern Africa ! And yet, withal, what faithful ardour drives him, that he should return again to woo his mate and rear his chicks in this grey twilight of our Northern night.

The path leaves the road and crosses the fields again. The shrill cry of the partridge comes up the breeze. A little while ago, leaving the beaten track, the foot stumbled into a cut thorn-bush on the open ground. Now where the grass is smooth and short the same accident happens again. We are in a land where the love of wild nature has left many a strange mark on character—a land in which respect for law still struggles unsuccessfully with the inborn belief that a man may take wild game and yet scorn to be a thief. The poacher loves these long even slopes as they will be later in the year, and the cut thorn-bushes have relation to his visits. The men walk them at night, two abreast and far apart, carrying a long, narrow net between them, slightly lifted in front and weighted behind. The birds lie on the open ground and do not rise. As soon as the net is over them they are doomed, and a whole covey may be captured at once. The thorn-bushes are the snares which wreck the net.

In the dim light mansions begin to loom out of the trees, and to take up the best positions on the higher grounds. The outskirts of the Metropolis have met us; just now, where no landmark showed the spot, the first boundary line was crossed—the line which marks the limits of the London Metropolitan Police area, a circle within which sleeps a population of millions. Under the oak copses the way winds. It is sheltered here from the north, and the air is warm and still. Hark! From the depths of the straggling thicket which skirts the wood there comes now a sound in which there is something curiously weird when heard for the first time and from a distance. It is a bird singing in the night. Clear, soft, and distinct, the notes rise and fall in the silence. It is the nightingale; this is a favourite haunt of the birds. It is surprising how far the sound travels; even after a quarter of a mile has been traversed in its direction it is still a considerable distance off. Similar sounds come now from the copses above, but the birds have each appropriated a situation; solitary they sit without changing position, each in continuous song throughout the night. It is the male bird which thus sings to the female as she sits on the nest. It is only a few steps from the thicket at last, and the songster cannot be more than twenty yards off. You do not wonder now at the estimate of the extraordinary quality of the bird's song, nor that it should have stirred the tongues of men to strophes in many languages. Full, rich, and liquid, the notes fall with a strange loudness into the still night. Yet it is not so much the form of the song itself which is remarkable as the passion with which it seems to thrill. Sweet, sw-e-e-t, sw-e-e-e-t—lower and tenderer the long-drawn-out notes come, the last of the series prolonged till the air vibrates as if a wire had been struck, and the solitary singer seems almost to choke with the overmastering intensity of feeling in the final effort. The stars shine through the feathery branches of the silver birches as you listen; the hoarse bay of the watch-dog still comes at intervals on the breeze; far down the valley burns the red eye of the railway signal; in the distance a coal-train is slowly panting southward, a pillar of fire seeming to precede it when the white light from the engine fire shines upon the steam: but the bird still sings on and on. It is lost in a world to which you have no key; it has not changed its position nor ceased its song since sunset, and it will be singing still with the dawn. Strange infinity of nature! Thus must its kind have sung here while the name of England was yet unfashioned on men's lips, and it was still a pathless wood to the northern Thames. Thus do the birds sing still on the fringes of modern Babylon, oblivious and indifferent to all that men consider the vast import of the seething life beyond.

The nesting season, when the birds sing, is drawing to a close. As the road winds near the copses the voices of other nightingales are heard, but they are not nearly so numerous as a few years ago. The birds are slowly retiring before the growth of the Metropolis. The writer's experience must have been that of many a Londoner in the outer zone. He has heard the bird from his bedroom window at night for a season; then the builder has come, its favourite grove or thicket has been cut down, and it has flown farther out, to return no more. The nightingales begin song here by the end of April, and they are almost silent by the end of June. They do not migrate till much later, and they continue year after year to frequent a locality until driven away; for, like the swallow, the same nightingale returns each year, faithful to its old haunts. The nightingale is not the very shy bird it is often supposed to be; although it usually keeps in the depths of its thicket, it may be easily seen moving about in the daytime. It sings then also, but its song is usually not continuous as at night.

The opal light in the north-east is spreading to the zenith. The path is through the fields again—another of those public footways which render England dear to the lover of nature. Although it is yet an hour and a half to sunrise, a red tinge is on the horizon, but everything is still ghostly and indistinct. Flip, flip!—a pair of larks flutter up from under the feet in the half light; they do not rise skyward, but they are already on the alert waiting to welcome the dawn. Hark! There is the first songster away on the right, the herald of the approaching day. This ridge is the last wrinkle of the chalk downs, the land which the larks love; from the next we shall overlook the outer rim of the great clay basin on which the Metropolis is built, and London will have straggled to our feet. A large grey bird, slimmer than a pigeon, sails out of the elms by the wayside into the morning twilight. It is the restless cuckoo, already astir. She does not call—it is too early. Besides, she has grown silent; the purpose of her strange, feckless life here is spent; a fortnight more, and her voice will no longer be heard in the land. The chorus of larks grows louder in the growing light. Already the southern slopes of London are in sight, shadowy and indistinct in outline, yet with a clearness rarely seen, and peculiar only to the smokeless summer dawn. Away still on the horizon runs the inner rim of the London basin, the line along which rise the heights of Richmond, Wimbledon, Sydenham, and Blackheath. Not so long ago, and its southern limit was still a wooded solitude; now the life of London has flowed far over its crest to the south, west, and east.

The bats are still wheeling in the streets of Croydon; a railway signalman swinging a red lamp crosses the way in front, and passes homeward; two men carrying lanterns and searching the ground pass down a yet unfinished side street. They are looking for the water-valves; this is the hour at which they can try the water in the new-laid connections with least fear of protest from the sleeping householder. Through the deserted roadways and sleeping squares the way mounts to the hill on which the water-tower stands. No other footsteps have broken the silence. Our janitor has kept his promise, and the key grates in the lock in a moment. Up we go the many steps—almost in the dark, it seems, for it is still nearly an hour to sunrise—and then out into the open at the top.

It is a strange world, dim and silent, which unrolls

itself before the eye here. There are in many ways few aspects of life more impressive than the awakening of nature on the fringes of a great city, and there are not many points of vantage better than this. Far below, the rows of houses and streets spread away on every side, the southern outskirts of the great circle, twenty miles across, which London occupies. Away to the north, farther in, though still only in the outer zone, rises the last ridge which shuts in the Thames valley; on its crest the gaunt glass structure of the Crystal Palace sits darkly on the horizon. Behind, to the south, stretch the downs we have traversed in the night. Between lies a great suburban land of brick buildings, new for the most part, here ranged in great solid blocks deep and wide, there straggling loosely apart. Everywhere between rise tall trees, now dark in their full summer foliage, the last survivors of that great North Wood in which, down almost into recent times, the charcoal-burners plied their trade—the North Wood which still gives its name to the district of Norwood, and which was so called to distinguish it from the other great wood, the Southern Weald, which stretched through Kent, Surrey, and Sussex. It is a fair land still, as it sleeps now under a cloudless sky out of which the stars have not yet faded, a battle-field withal—a land upon which the invading Celt and Roman and Saxon has in turn left his hand, it is true—but a battlefield, most of all, where nature fights year after year a losing stand against the blighting and despoiling forces of civilization.

Hark! There comes now the first sound from below. It is a thrush tuning for the opening symphony. After a few tentative notes it bursts into full song. Cherry-dew, cherry-dew! Be-quick, be-quick! Strangely clear and distinct, the full notes ring out in the still morning. Soon it is joined by another, and in a moment another and another have answered from the high elms around. The volume of sound continues to grow, but as yet it is only the thrushes which greet the dawn. Soon there reaches the ear a faint, harsh murmur; now it is louder, and soon it swells into a hoarse din. It is as if a great army of workmen had suddenly begun to labour below, and the harsh chip and fret of countless iron tools rose upward in blended discord. It is the multitudinous voice of the house-sparrow. He rears three families in the year, and he has begun his day's work of eighteen hours. He it is who, alone of wild birds, can regard the nineteenth century as an era of unexampled prosperity. He has multiplied in incredible numbers with the growth of towns. Nay, more: following the Anglo-Saxon, he has spread with the extending race to the ends of the world, till over two continents, with a certain appropriate inaccuracy, he is known and banned as the English sparrow. From the lower shrubs of the private gardens the rich, mellow note of the blackbird begins now to blend with the others. Louder and louder swells the chorus of voices, as the finches, robins, and other small birds join in at last. It is a strange harmony—one which is seldom heard by the sleeping world. The strangest feature is, indeed, the almost complete absence of any human sound; save for the occasional scream of the whistle of a locomotive shunting on the distant line, all but the voices of the birds is silent.

Round the tower the bats are still hawking.

From below there reaches up a familiar twitter. It comes from a line of swallows which stand huddled up after the night on the paling, their white breasts showing in marked contrast to the black-painted fence. One takes wing now, at last, to begin that long chase after flying insects which the bats have not yet abandoned. Thus do the fringes of the night overlap the coming day.

As the light grows, the features of the land open out. One does not wonder here why the migratory wild birds come to us in the far North-west in such numbers. Why should they linger amid the barren larch plantations and the *petite culture* of the Continent? Where else, despite the growth of the towns, has the country been preserved so unchanged as in England? To the right stretch the natural woods and copses in the direction of Chiselhurst; nearer at hand lie the Addington hills and the splendid wooded lands of the manor of Croydon. Away to the left roll the level plains toward Windsor, the great trees so thickly strewn over the land as almost to give it the appearance of a thickly wooded country —trees which rise unkempt in the free air of heaven, with limbs unlopped, in all their natural beauty. To the south stretches the open land, the commons of Epsom and Banstead, and the range of the North Downs, with the little village of Purley, associated with the fame of Horne Tooke, sleeping on the edge. It is all little changed since the days of the author of the "Diversions," always and except for the vast growth of London. What would the eccentric parson and politician have thought of the age if he had lived to see the Metropolis almost at his doors, and all that the whirligig of time had brought with it? Would he have thought any better now of the crime which split the Anglo-Saxon peoples in two, or of his countrymen who fined and imprisoned him for opening a subscription for the widows and orphans of the Americans "murdered by the King's troops at Lexington and Concord"?

The rooks are spreading out across the sky as they sail from their nests to the distant pastures. As the light ripens, the view enlarges of greater London stretching away to the north. Like the arms of a great octopus, its fringes strike far into the open land. Farther in, caught between them, rises bravely many a pleasant grove; parks, open spaces, and even fields gleam a fitful green among the bricks in the morning light—but surrounded all; doomed, injected morsels waiting to be digested at leisure, to serve the strenuous purposes of another life. And yet only the outer suburban zone is visible here—a land of beauty without refinement, of wealth without distinction; a land of groves and spires and villas hedged round with reformatories, schools, and asylums. And everywhere, from horizon to horizon, the unfinished brick and timber of the builder, emblems of the ever-rising flood, of a movement of which the springs are at the ends of the earth, of a life which takes toll of every land under heaven.

Now at last, away in the north-east, the fiery red rim of the sun shows above the horizon. There has been no gorgeous preparatory display, no massing of shades and colours for the opening ceremony. With scarce an anticipatory flush he rises full into a grey, expressionless sky, and a moment afterward disappears into a bank of fog which hangs on the horizon over the Essex marshes. A fitting tribute, perhaps, to the race and clime. For he has risen over the first meridian, over the mother city of the Northern Vikings. It is from here that the nations have learned to count their distance. It is from here that they measure his course in his race round the trackless seas.

VII

THE HAUNTS OF COOT AND HERON

IT is early morning. A grey steam rises from the surface of the water through which we wade, hidden in the long green sedge which, although it is only the growth of a few weeks, already meets overhead. Later on in the day the long May sunshine will warm the shallows, but now the keen night of the early sunmmer has chilled the air, and the water appears to break into vapour with every step. Wade gently, for in these lonely haunts of coot and heron we at this hour steal upon Nature in one of her most private moods.

The soft peaty bottom has changed, and we have reached an opening among the reeds and flags. It is a pool half a dozen yards across, strewn deeply with clear gravel below, and circled almost completely round with rank sedge, which hides the view on all sides. It is a miniature delta in process of formation by the tiny rivulet whose waters, the murmur of which is just heard in the distance, here lose themselves in the lake. There is a splash and a deep swirl as the foot grates upon the gravel ; another and another. We have disturbed the pike, which had come here to feed in the night. The practised eye can just distinguish the flat head and grey-green outline of one which has not yet moved and which is still lurking in the deeper water. Every living thing of suitable size is prey to these hungry jaws. Many a pair of bright eyes belonging to a little ball of black down has set out to cross this treacherous pool but has never reached the other side. A dull splash, an eddy in the water, and the owner has disappeared.

The water is nearly over our deep wading boots. Drive your stick into the sand and listen as you lean. We are in a land three hours from the nearest railway station, and ten hours even then from London. Over the wide expanse of water in front which you cannot see, comes the harsh sound of many voices. It is the screaming of the water-fowl, but so subdued by distance that it almost forms part of the silence. A few notes with long spaces between from the wood-pigeons in the wood behind sound so clear and distinct that they render the early morning stillness in the interval more marked. A loud croak close at hand causes you to turn. There is a faint rustle and a glint of white and bright red amongst the green, as a moor-hen, jerking her tail impatiently, looks out from the reeds. She wants to cross to the other side, and finds herself intercepted. Her movements betray the mother, and it is easy to tell that her nest is not a dozen yards off. She strikes out now into the deep jumps clear on to her feet, and races with outstretched neck to the other side, apparently on the surface of the water, but in reality supported by the broad leaves of the water-lilies—which the country folk here call drowning lilies, from the belief that the long rope-like stems on which the leaves rise from the deep bottom will entangle and drown the strongest swimmer.

Motionless and without a sound as we stand, the knowledge of our presence appears to have travelled mysteriously on every side. All wild nature has an instinctive knowledge of the proximity of man. An air of suspicion and caution seems to spread all round; the ear strains after faint rustlings and plashings, which it fails to distinguish. Hark! a peculiar note breaks the silence. It comes from the open water in front. It can be compared to nothing else than the sound caused by dragging a jagged piece of metal over a slate. It comes closer in, till at last the outline of a bird, smaller in size than a full grown duck, can be distinguished through the reeds. It turns its face towards us now—a face perfectly white as seen against the deep black plumage of the rest of the bird. We do not move, and the creature presently works itself into a royal rage; it sails up and down furiously, even stopping at times to stamp in the water with its feet, like a child in a passion. Foolish creature! only a bird with a long line of inedible ancestors like the common coot would venture into such tantrums in the presence of the lord of creation; even a tyro with a gun could not miss it now. But you remember that it is the breeding season, and many and strange are the instincts with which Nature has endowed her creatures at this time.

Here in deep water amongst the sedge is the cause of all these angry passions. It is a curious structure to look at, but, when the adaptations of means to an end are considered, one which must exact admiration. A huge submerged heap of water-soaked rushes and reeds is surmounted by a deep layer of the same materials, warm and dry, and lined inside with smaller chops and fibres. The deep hollow in the centre holds nine large eggs speckled with black on a grey ground colour. The eggs are quite warm, for the bird has been sitting during the night, and she has doubtless only just slipped into the water.

We have reached a spot now where the rushes grow thinner, and where the reeds, which abhor a gravelly bottom, are found only in patches. The fat perch, which always share these inland waters with the pike, move lazily into deeper water as we advance. You notice a wild movement amongst the sedge in front. Something is beating its way before us. It is not a dozen yards off now, and the bulrushes and long reeds are violently agitated as it moves along. It cannot be a dog, it is not an otter, and no fish would cause such a disturbance. Nor is there any bird which would exactly suggest such a movement. We press forward, and the distance is lessened : it travels slowly. Now we catch sight of something brown moving. Another stride, and the cause of the agitation is revealed. It is a wild duck—the mother bird—pressing her way through the sedge ; not alone, however, but closely followed by eight or nine recently hatched little ones, the latter so closely packed together as they swim that they seem to move through the water like a solid bank of dark brown fur. We are seen. Now you witness one of the most curious sights in nature. The mother at once abandons the efforts she has been so far making to glide away with her charges without attracting attention. She jumps clean out of the water—not, however, to fly away, for alas! she falls back again heavily and apparently helplessly a yard away, painfully flapping a broken wing. You are conscious that her brood scattered cheeping in all directions as she rose from the water, but your eye is back again in an instant, and lo! there is not one to be seen, and the cheeping has ceased almost as soon as it began. There is no longer, you notice, the least tell-tale tremor in the sedge to show where they have gone. The old bird continues her antics with the broken wing. You have seen the sportsman's victim acted in real life, and you know that even a professional medium could not more thoroughly abandon herself to her part. You may watch her at leisure, for she does not in her sad plight seem able to get under cover quickly. Her callow offspring you will see no more.

Often as the writer has witnessed this little scene enacted, it has never ceased to fill him with surprise. He has often set himself to watch how it is done, but he has never been successful. He has surprised the mother bird in a patch of sedge so short that there has been practically no cover, and yet he has not seen the young birds dive—for dive they must—the antics of the mother bird always, in spite of self, engaging the first glance of the eye. Nor, stranger still, has he ever seen the little divers emerge again, although of course they cannot remain permanently under water. Sometimes, but very rarely, you will come across a little downy body in the water with legs hanging motionless downwards, and only the little beak projecting just above the water, looking like the broken-off end of a last year's bulrush. But if this is how the birds hide themselves, how do they render their otherwise buoyant little bodies suddenly heavy enough to enable them to retain such a position under water?

As we continue to move through the sedge you notice that the unusual exertion is having a wonderfully curative effect on the broken wing of the mother. She is already taking short flights with it, although still occasionally flopping back heavily into the water. As you look she sits up and flaps both wings airily enough. Now she springs into the air, and, wheeling several times nimbly overhead, actually takes her departure altogether, with a series of wild, derisive quacks as a parting salute. You feel somehow as if you had not got the best of the encounter, and that you have been treated throughout as creature of inferior intelligence.

Here, where the ground has become spongy again the green mare's-tail spikes grow thickly together near the edge of the water. A spot where a number of spikes have been brought together at the top attracts attention, and, following the tell-tale appearance downwards, you are moved to admiration by the sight below. A snugly placed ball of dry warm rushes and grass has been put together. In the hollow are packed eleven eggs, considerably smaller than those of the coot, and speckled with rather larger dark-brown spots. It is the nest of a moor-hen. One of the eggs is already fractured by the little chick within, and within a few hours the whole brood will be in the water.

We must leave the water now, for the sedge has suddenly ceased, and it is too deep to wade here. We sit down on the grey shingle, worn smooth by the winter floods, but now high and dry in the morning sunshine. See, slowly over the wood comes the monarch of these waters, his wings flapping slowly and deliberately while his keen eye examines the landscape. The first heron has come to break his fast, and he will soon be followed by others. Slowly lowering in flight, he wheels over the long shore line, taking it first from the lake and then from the land side, to see that all is safe. He has seen us long before we were aware of his approach, and he will give our neighbourhood a wide berth. No bird is more wary than the heron ; he is choosing his position with all the caution of a general. Three points of vantage it must combine to suit him : it must give him a view of all the approaches ; it must at the same time screen him from view ; and it must be a good fishing ground. He alights at length, and you see how well the position has been selected. Were it not that the eye has kept the grey plumage in sight the whole time, it would not now readily distinguish it, partly screened as it is by the sedge, which you notice is at the spot chosen short enough to allow him, when the long neck is fully extended, to command a view of the whole neighbourhood.

He stands where the water reaches half-way up his long legs, and with a pocket glass it is interesting to watch his movements. He does not walk about after his prey, as the waders often do—as you may see the storks do on the flats in Holland or in the Rhine marshes below Mayence. The heron sometimes wades—when necessary—but here he stands motionless as a sentinel, occasionally bending his head slowly downwards or on one side, to see better what is going on. Nearly everything that moves in the water is prey to him, but here he often aims at big game, and even the large perch and hungry jack do not always escape him. Deceived by his motionless attitude, and doubtless taking it as part of the fixed and natural order of things that such a colossus should bestride their narrow world, they swim between his legs to their doom. The sharp, powerful, bayonet-like beak is a most formidable weapon. The fish is struck in the water, caught in the bill and held aloft. If it is a small one it is swallowed forthwith, with a toss of the head; if the capture is too large to swallow, the bird's action is different. In these waters, at certain times, particularly during the spawning time, the heron captures many larger fish; the prey in such cases is carried to land and the eyes are picked out; as a rule the rest of the body is not eaten.

A beautiful bird is the heron at close quarters. The rich yet delicate grey colour of the greater part of the plumage, the black breast with its white patch in front, the crest of black plumes on the head, and the pendent breast feathers of the adult, all combine to give a peculiar air of distinction to the bird. It is for this reason that it is so often shot at, and in consequence it has become shy and wary to an extraordinary degree. It is in many districts a feat requiring no small skill and patience to shoot a heron. He knows a gun in the hand at a distance as surely as a rook. He keeps out of range as carefully as a curlew, and he has to be stalked to windward with nearly as much caution and cunning as a royal stag.

A few of the birds breed not far from here. Behind the spot on which we are seated the land slopes gradually upwards till it rises into a bold, craggy ridge two miles away. Then it drops down again, not now gradually, but suddenly and almost precipitously, forming a deep, narrow valley, shut in on the other side by a similar formation. The silver firs which have grown here have been well protected, and they rise to a great height, as they sometimes do in the lower valleys of the Alps in similar situations ; and here, too, they are straight and shapely from the ground upwards. On the tops of the highest trees there are about half a dozen nests ; and the sight as seen across the valley from either side in the breeding season is not soon forgotten. The tall, gaunt forms of the birds, perched like sentinels on the nests or branches, or occasionally arriving and taking flight amid a chorus of solemn croaks, the young in the great flat nests—not seated as young birds usually are, but also standing, the stilt-like legs supporting bodies covered with long loose down—all combine to give a peculiar air of old-world stateliness to the picture, which is very striking. No wonder the heron was a bird held in favour by our ancestors ! The heron begins to sit about the end of March, and the nest is always built on a tree of considerable altitude, and generally near the top ; it is constructed of twigs principally. From three to five eggs, of a blue-green colour, are laid, and the young remain long in the nest after they are hatched out.

Two more of the birds come now over the hill. They fly close together, but separate as they descend, closely examining the while the lake-shore in wide leisurely circles of flight. They have chosen for their fishing-ground the end of the lake where the overflow water spreads out over a vast level expanse of peaty ground, converting it into an immense swamp where no boat can penetrate. If we would follow, we must be careful. The water is not deep ; but in long past days, probably when the sluggish river at the end of the valley was less choked with weeds, the unremembered inhabitants cut peat where the water now covers, and their deep pits, overgrown with weeds and nearly obliterated, still yawn treacherous in the way of the unwary. This part is the resort of great numbers of water-fowl, and in the winter season it is visited by many migrant species. The wild ducks in particular come here, and many pairs, attracted by the seclusion and abundant cover, remain to breed. Great bosses of coarse grass, which rise high and dry out of the water, mark where the lake ends and the shallower water begins. Closer and closer these miniature islands get, and the long rough sedge blades, which at certain angles cut like a knife, stream from off them into the water, forming cover through which the wild fowl have worn openings like those which the rabbit makes through the hedgerows. Farther back the reeds rise higher, and there are deep pools of open water, and then again more sedgy islands on which the black willows eke out an amphibious existence, maintaining an unequal struggle against their many aquatic rivals.

One may wade for long here and see little or nothing of the bird-life with which the place abounds. It is necessary to lean silently against one of the clumps of sedge grass and wait for it to venture out into the stillness. We may step out of the water here. The long coarse grass is virgin to human foot, and, mixed with the short willow twigs which struggle through, it forms excellent cover. We must search here. The scrutiny is close; every tuft of tangled sedge grass is explored, yet without finding anything. Just as we are about to give up hope we are successful. Not where the cover is thickest, but, after all, where the nest is best hidden, we find it. The wild duck is a wary bird in all her movements, but most of all is she careful in risking the safety of her nest or her young. There are nine eggs, large, warm, and nearly hatched, and of a very pale blue colour; for the eggs of the mallard have not nearly so much green in them as those of her domestic relative. The nest is made of dry grass and is lightly lined with feathers. You have to acknowledge that you are not superior to a slight tightening of the throat at the sight of this nestful of the large eggs of this shy bird. Yet you remember that they are perfectly useless to you. You would not, moreover, willingly remove one of them. It is the fascination of meeting wild nature face to face, and wresting her spoils from her fairly, which stirs the old Viking blood. The strain is there; it breeds pirates or poachers, merchant adventurers or world-builders—according to circumstances.

A movement overhead causes you to look up. A great black bird, almost as large as a goose, but of slimmer build, is circling in wide descending circles. It is evidently going to alight in the water, but at a distance, for it will keep carefully out of range. The bird is a cormorant. Individuals of the family often come here to fish in certain states of the weather, and they prefer the still early morning.

Looking westward, we are not a dozen miles from the sea, and the birds are equally at home in fresh and salt water. It alights at last well out from the shore, where the limestone crag rises highest out of the water. The bird for the greater part of a minute remains motionless on the surface, with neck erect, surveying its surroundings before getting to work. Now it has dived. The still black water into which it has gone down is said by the simple country folk to be bottomless. You know it is not, but you know also that it is deep—sixty feet at least—and the bird will reach the bottom. The moments go by, and it does not reappear; a man, you think, could not hold his breath so long; and yet it does not return. At last, after what appears a surprisingly long interval, it emerges. The long neck is not now erect, but is held in position as if the bird were panting after a supreme effort. Now it raises its head again, and you see that it has not returned empty. It has got something in its beak, something which twists and knots itself about the outstretched neck. The bird jerks and tosses its head in the effort to swallow it; it is an eel, you see plainly, an eel which must have been brought up from the depths below—one, too, which is by no means resigned to its fate, and which is resisting to the last. With no inconsiderable effort the bird at last succeeds in swallowing the prey, after which it sits for an interval quietly on the water resting after its exertion.

Now it is down again, but it soon returns, this time empty. It is off again, and after another prolonged absence it returns with a second eel, and the same struggle between the fish and its captor is gone through before the prey is swallowed. There can be no doubt as to the skill of the bird in fishing, and it is hard to imagine how it has acquired such extraordinary keenness of vision and swiftness of movement under water as to be thus able to outmanœuvre the fish in their own element. One can readily imagine how such exceptional powers should have suggested to various peoples the idea of utilizing the bird in the service of man. The mate is now seen circling overhead, but it does not alight, and the bird in the water takes flight at length, the two wheeling round and round in wide circles until they reach a considerable altitude, when they sail out of sight seaward.

As we return by the lake shore the rabbits are feeding close up to their burrows, and the rooks have long since scattered to the distant pastures. The coots are taking their families out on the surface of the lake, leaving the friendly shelter of the sedge; and the queer little black balls swim obediently behind, scarcely venturing out of line even after the most tempting morsels. Foolish mothers! Not thus does the mallard risk her offspring in the open. They look safe enough, you think, but they are not really so. Ah! to be a boy was to look upon every young water-fowl which took the open water as delivered into your hands. It was only a question of time and dexterity to effect their capture, barehanded and without boat or net.

It was not so very difficult. Only two qualifications were necessary. You must be a boy, and a good swimmer—especially the first. A tyro might capture one, or even a pair; but to secure a whole brood of the nimble little swimmers at a single stretch was a feat which justly entitles to distinction. How was it done ? To be successful it was necessary to take the water bravely, like a retriever, and with a strong breast stroke, which soon lessened the distance between you and your quarry. As you approached, the struggle in the mind of the mother bird was always ludicrous. She was torn between the two great forces which move nature's world—self-interest and parental instinct. First she would and then she wouldn't leave them. But she always did, after all. It is a very pretty sight. Down all the little swimmers go immediately the old bird flies away ; they feel they are left to their own resources now, and they scatter in all directions as they dive. Now, if you are to return successful, your strategy begins. Slowly swimming in the direction in which you have come, you wait. Bravely the little divers act their part, long do they stay down, and far do they travel before they emerge. They come up at last ; and, singling out one from the rest, you again pursue it. Down it goes once more, and it always swims under water in a straight line away from you —a fatal mistake. This time it stays below a shorter interval ; and a few more trials and you overtake it, and it submits to be caught. The little black leg must be held gently in the mouth, and the little owner floats comfortably in the water without struggling while you proceed to capture the others one after another in similar fashion. Then, having covered over two miles in the water, panting, flushed and triumphant, with five little cheepers, frightened but not hurt, streaming from your mouth, you swim to land under the envious and admiring eyes of your equals. You may afterwards worst your fellows in competitive examinations; you may climb up the ladder of life two steps at a time; you may woo and wed the woman you love; you may even publish your first book and read the reviews of it. But never will you be any happier than that.

The sun has warmed the shingle when we return to it again. The steam appears to have ceased to rise from the water: it has only become invisible in the warmer air. The wood-pigeons have left off calling to each other, and are flitting to and fro in the neighbourhood of their nests. The perch are already chasing their own offspring near the surface, the little fugitives at times jumping clear out of the water to escape capture, and falling back again with a sound as if a handful of fine gravel had been thrown into the deep. Hark! over the water comes the only sound from the outer world which reaches these solitudes—the faint tinkle of the morning bell calling the toilers to work in the distant quarry in the hills. The long, still, early summer morning, when all wild nature lives and moves, is waning at last. The day has begun.

VIII

CONCERNING THE CUCKOO

THE cuckoo has the distinction of being one of the best known and least understood of our British birds. If all the literature which this strange bird has inspired were collected together it would form a small library in itself. Yet there is scarcely a point in connection with its curious life-history which is not from time to time made the subject of question and even contradiction by competent observers. The brief, mysterious visits to our shores, the sudden appearance everywhere in the early spring, and disappearance equally sudden when the year has but reached its zenith, the shy and unsociable habits, and above all the legend which from time immemorial has attributed to the bird conduct both as a parent and a nestling so unnatural as to be almost without a parallel, all combine to give the cuckoo a place in popular imagination which no other bird can lay claim to.

When the month of April reaches its teens the cuckoo comes amongst us in the south of England. It goes north with the advancing year, and appears generally in Scotland about the beginning of May.

One of the first things which attract the attention of every observer of the habits of the bird is the manner in which it distributes itself over every variety of country in these annual invasions. Other migrants have their favourite haunts: the nightingale seeks the copses of the southern counties; the lark and plover the open moors; the swallow the pastures, open waters, and the haunts of men; the mud-flats, the deep woods, and the rocky places have each their special *habitués*. But the cuckoo is to be found nearly everywhere. It takes the woods of Hampshire as familiarly as the trim poplars of the Continent, and it spreads itself over hill, dale, and open country indiscriminately. The cuckoo is common round the fringes of London, apparently because of the presence of the numerous thickets in which it delights; but it remains where trees and even hedgerows fail, for it may be seen in the bare mountain-limestone country, with not a bush in sight, flying familiarly from stone to stone and making the rocks echo with its well-known call.

The cuckoo cannot properly be viewed from one standpoint. All its habits form part of a single study. Even this apparently incidental question of wide distribution and adaptation to diverse localities is probably intimately associated with the other unusual habits of the bird, and must be considered in connection with them.

Of the actual existence of the most widely reputed habit of the cuckoo, that which has led to the popular estimation of the bird as a monster of treachery and immorality, there can now be no possible doubt. The tradition respecting it is of great antiquity; but, unlike most traditions in natural history, it has been for long supported by observations numerous and authentic enough to satisfy the most exacting. Where the cuckoo is plentiful almost any painstaking observer will be able to find for himself the intruder's egg in the nest of one or other of the species of birds commonly made use of. The mother has been caught by many observers in the very act of foisting her offspring on her neighbours, and the young bird has been followed in every step of its adventurous career from the egg to the adult. Nothing in fact has been left undone necessary to satisfy the utmost scruples of anyone gifted with that sceptical bias in these matters which the pursuit of science is supposed to demand.

The eggs of the cuckoo have been found in the nests of nearly every species of bird in Great Britain and the Continent suitable for its purpose. The nests principally made use of in England are those of the meadow-pipit, hedge-sparrow, and pied-wagtail. In certain districts where the reed-warbler is common the nest of this bird is a great favourite, and the same may be said of the redstart. Although the range of choice which the cuckoo exercises is very wide it is a noteworthy fact that the bird nearly always chooses a nest belonging to a species the natural food of which is suitable to her own young. The foster-parent is thus nearly always insectivorous, the nests of birds which feed on vegetable substances being rarely used. Even the best regulated instinct, of course, sometimes errs, and the cuckoo's is no exception to the rule, the unnatural parent sometimes providing foster-parents equally unnatural for her young by occasionally depositing her eggs in nests such as those of the wood-pigeon and house-sparrow. But the instinct which leads the bird to choose the right nest is well marked despite these occasional lapses, and we shall have a word to say directly as to the manner in which it probably originated, in common with the cuckoo's other peculiar instincts.

Nothing connected with the cuckoo has given rise to so much discussion as the extraordinary character of its egg and the manner in which it is placed in the nest chosen to receive it. Every one who has collected birds' eggs, or indulged in the juvenile habit of birds'-nesting, or who has even gone so far as to take an intelligent interest in the dozen of new-laids ordered from the grocer's, must have noticed one rudimentary fact respecting the eggs of birds. The eggs of each species have certain marked characteristics which distinguish them from those of other birds: the common fowl's egg is white, the duck's pale blue, the thrush's speckled green, the skylark's dark brown. The eggs of each kind of bird also vary but little in size. Now, strange to say, the cuckoo's egg is a marked exception to this almost invariable rule. The eggs of the cuckoo have no particular colour. They have been found green, grey blue, grey-mottled, green-mottled, and pure white. Neither have they any particular size. They vary in the most puzzling fashion, from the size of a skylark's egg to almost that of a pigeon's. Few of the authorities on the subject can agree even as to what the average size should be. For instance, two of the best known, to whom I refer at random, state the size of the cuckoo's egg to be respectively $\frac{35}{40}$ inch by $\frac{27}{40}$ inch, and 1 inch to 1·8 inch by ·75 inch to ·61 inch—a sufficiently wide difference almost to suggest a doubt whether they are really speaking of the egg of the same bird.

Closely associated with this question of the unusual variation in the appearance and size of the cuckoo's egg is that of the character of the nest in which it is laid. Formerly, before the habits of the cuckoo had been made the subject of such close study, the prevailing idea was that the bird sought out a convenient nest, apparently at random, and laid an egg in it in the absence of the owner. More systematic observation has, however, revealed that the cuckoo's meanness has more method in it, and method too which is apparently most skilfully devised to attain certain ends.

A great number of authentic observations, made in a variety of places, appear to have established it as a fact beyond doubt that the eggs of the cuckoo are as a rule deposited in the nests of birds whose eggs approximate both in size and appearance to the strange egg placed among them. The view previously held that the cuckoo actually laid her egg in the chosen nest has been considerably modified by observations both in this country and on the Continent. It seems still probable that the cuckoo sometimes lays in the nest, particularly when it is open and conveniently situated, but the general habit of the bird would appear to be to lay her egg on the ground first and then to take it in her bill and deposit it in the selected nest.

This method of depositing the egg, taken in connection with the acknowledged fact of the variability of the cuckoo's eggs and their general approximation in appearance to the eggs with which they are placed, has led to the formulation of two theories on the subject, both of which are steadfastly held to by their advocates.

According to the first view, the cuckoo, having chosen the nest of a bird in which she is about to lay, has the extraordinary power of being able to control at will the appearance of her egg. She is supposed to be influenced in some unknown way by the surroundings or the appearance of the eggs already in the nest, and so proceeds to produce an egg resembling those of the foster-parent. The other theory credits the bird with scarcely less originality, though with more shrewdness. According to the second view, having laid an egg on the ground, she takes a kind of mental inventory of its appearance, and then proceeds to deposit it in the nest of the bird whose egg it resembles.

Without staying at this point to discuss these and other theories which have been put forward to account for the curious fact that there is usually a general resemblance between the cuckoo's egg and those of the widely different species of birds with which it is found, it may be mentioned that it is likely that the cuckoo often lays on the ground without the intention of placing the egg in any nest, and even possibly occasionally settles all question of its destination by quietly making a meal of it. The bird seems in some manner to have obtained the reputation of an egg-sucker, but whether on the strength of reliable evidence or not it seems hard to say. I recently caught a cuckoo in the act of laying on the ground in somewhat peculiar circumstances which have some bearing on this point. Returning across Wimbledon Common about dusk, on passing a thicket in one of the retired corners, I saw a cuckoo, which was calling, flying low and in a peculiar way over the bracken. I stopped and watched the bird, and saw it alight down suddenly out of sight in a meaningful way. Hastening up to the place, I came upon two cuckoos in a dry open space among the ferns, one of them apparently in the act of depositing an egg. Both birds flew awkwardly away on my approach, and I took possession of the egg, which was quite warm. Most careful search was made all round the spot within a considerable radius, in the hope of finding the nest of some small bird for which the egg might have been intended, but no nest of any kind was found. A point which, however, seems worthy of remark is that on afterwards returning to the spot where the egg was picked up I found the broken remains of a similar egg which had apparently been sucked. The conclusion which presented itself to my mind at the time was that the bird had not intended to deposit the egg in any nest. She had probably laid in the same spot before, and had either feasted on the first egg herself or had left it, and it had been found and sucked by some animal. The second egg would most probably have suffered the same fate.

It is somewhat strange to find that there is still a difference of opinion as regards the behaviour of the nestling cuckoo towards the young of its foster-parents. That the presence of the young bird is fatal to the other birds in the nest is universally conceded, but that the interloper actually and deliberately throws out the rightful owners of the nest, in order to monopolize the whole of the parental care, is still questioned by writers of authority.

The presence in the nest of a bird so greedy as the young cuckoo, and usually so much bigger than its fellows, would, it is urged, in any case bring about the death of the latter, and without it being necessary to assume any *malice prepense* on the part of the young cuckoo.

There seem, however, to be no grounds for acquitting the bird of the charge of deliberately and intentionally causing the death of its fellow-nestlings. Not only is it certain that the young cuckoo ejects the other birds from the nest, but it would appear to be also true that several details of its anatomical structure, and even the temper and disposition of the bird during the first few days of its life, have been acquired for the special purpose of executing its murderous work as swiftly and efficiently as possible. Soon after the young cuckoo is hatched out it exhibits an extraordinarily irritable and restless disposition. It keeps on beating its stumps of wings, it tries to get underneath anything that may be placed in the nest. Anyone may see by a simple experiment how the bird regards itself in relation to all comers. Not only will it put out the other occupants, but it will throw out pieces of wood, lumps of earth, the eggs of other birds, or anything of the kind which may be placed by the observer in the nest. The other nestlings are usually disposed of at once—that is to say, during the first or second day—and any eggs that may still remain unhatched in the nest are put over the side at the same time.

The surprising and exceptional nature of this phenomenon, and in some measure also the difficulty of accepting the explanation usually given of the origin of the instinct in the young bird, must be held to account for the disposition shown to accept accounts of it with reserve. One of the most graphic sketches of the occurrence by an eye-witness is that in Mr. Gould's "Birds of Great Britain." The account by Mrs. Blackburn, who watched the movements of the young cuckoo, is full of interest. The nest under observation was that of the common meadow-pipet, and it had at first two eggs in it besides that of the cuckoo. "At one visit," continues Mrs. Blackburn, "the pipets were found to be hatched, but not the cuckoo. At the next visit, which was after an interval of forty-eight hours, we found the young cuckoo alone in the nest, and both the young pipets lying down the bank, about ten inches from the margin of the nest, but quite lively after being warmed in the hand. They were replaced in the nest beside the cuckoo, which struggled about until it got its back under one of them, when it climbed backwards directly up the open side of the nest, and hitched the pipet from its back on to the edge. It then stood quite upright on its legs, which were straddled wide apart, with the claws firmly fixed half-way down the inside of the nest, among the interlacing fibres of which the nest was woven, and, stretching its wings apart and backwards, it elbowed the pipet fairly over the margin so far that its struggles took it down the bank instead of back into the nest. After this the cuckoo stood a minute or two, feeling back with its wings, as if to make sure that the pipet was fairly overboard, and then subsided into the bottom of the nest." The ejected bird was replaced, but on again visiting the nest on the following morning both pipets were found dead out of the nest. Mrs. Blackburn continues: "The cuckoo was perfectly naked, without the vestige of a feather, or even a hint of future feathers; its eyes were not yet opened, and its neck seemed too weak to support the weight of its head. . . . The most singular thing of all was the direct purpose with which the blind little monster made for the open side of the nest, the only part where it could throw its burthen down the bank. I think all the spectators felt the sort of horror and awe at the apparent inadequacy of the creature's intelligence to its acts that one might have felt at seeing a toothless hag raise a ghost by an incantation. It was horribly uncanny and gruesome!"

In a nest which the writer had under observation, the little cuckoo had put one of four hedge-sparrow's eggs over the side on the second day of its existence. In another nest under observation at the same time, the young hedge-sparrows were hatched out so long before that soon after the cuckoo was hatched its nest mates were at least four times its size. But at this stage the young cuckoo was seen to put the hedge-sparrows, one after another, over the side of the nest till it was the sole occupant. No one who has not actually seen the process of ejection of the other young birds can fully realize the uncanniness and almost incredible purposiveness of the whole series of actions. At the time when the bird's instinct is at its maximum the young cuckoo is only a few days old. It is blind and naked, without the vestige of even the beginnings of a feather, so that it presents the very image of weakness and helplessness. Yet in such circumstances it sidles up to the other occupants of the nest, using as feelers the long and bare wing processes, which have an appearance strangely suggestive of the arms of an ape. Getting gradually under its fellow-nestling, it lifts it on to the flat back; then using the ape-like arms as props and the strong legs as levers, it partly raises and partly pushes the victim upwards, clambering backwards up the side of the nest. When it reaches the edge the victim is hitched over and the last scene of all almost takes one's breath away, for the blind little creature, before returning to the bottom of the nest, feels round as if to assure itself that the difficult business had been in all respects successfully accomplished. After its efforts the cuckoo appears completely exhausted. But it resumes its attempts when rested, and it will continue for days to eject any other birds or eggs that may be placed with it in the nest.

The number of the theories which have been put forward from time to time to account for the unusual habits of the cuckoo is legion. The instinct of the young bird is surprising enough in itself, but the disappearance of the parental instinct in the old bird, the habit of depositing its eggs in the nests of other birds, the extraordinary variability of the egg and the character of the nest in which it is placed, appear to be quite as difficult to explain.

One of the theories respecting the cuckoo which has received general support is that the bird's parasitic habits are the natural result of the character of its food. This matter has an interesting aspect. Those who are familiar with the natural selection theories of Darwin, Wallace, and Lubbock will know that certain hairy caterpillars are supposed to have acquired their striking appearance as a protection from birds. Strange to say, however, it is these caterpillars of the hairy kind, which other birds leave, which form the staple food of the cuckoo. The bird, by universal consent, is enormously greedy, and it devours great quantities of them. Now, it is pointed out that, as the supply of this food soon fails, the cuckoo is obliged to migrate so early that it would not have time to take upon itself the cares of maternity, and so it has acquired the convenient habit of placing its offspring out to nurse. Unfortunately, however, for this theory, there are several difficulties in the way of accepting it as it stands. The habit is said to be found in the Indian species, which do not migrate. The old birds leave us in July and August, but the young remain a month or six weeks longer ; and if they can find food, why not the old birds ?

Another theory which has its supporters is that the parasitic habit is the result of a peculiarity in the manner in which the cuckoo's eggs are laid. It is now well known that the bird does not deposit her eggs rapidly like most birds, but that an interval of four or five or even eight days intervenes between them. Hence it is said that the cuckoo evidently could not utilize a nest of her own, for the first eggs would be addled or hatched before the last were laid. There are difficulties in the way of this theory too. There are other birds who lay their eggs in the manner of the cuckoo, but without having acquired its parasitic habits. Irregularity in this respect exists doubtless to some extent in many kinds of birds, and in some to a considerable degree. Mr. Cones says of the American species (*Coccygus*) that the nests commonly contain young by the time the last egg is laid. The evidence would seem to suggest that this habit has been developed in the cuckoo rather as the result of its other habits than as the cause of them.

There is another theory which has received the adherence of many persons of weight. One of the strangest of many unusual facts regarding the cuckoo is the proportion of the sexes. The males greatly outnumber the females. The males have been estimated at ten to each female, and by some observers as high as fifteen to one; even the most moderate estimates do not place the proportion at less than five to one. The theorists who find in this fact the cause of the peculiar habits of the birds are, however, not agreed among themselves as to how it has operated. Some regard it as precluding the cuckoo from mating in the ordinary way, and so from building a nest and rearing her young. Others regard the temperament of the bird as a kind of physiological accompaniment of the relationship of the sexes, but on grounds which seem rather unsatisfactory, if not obscure.

It is probable that any satisfactory explanation of the unusual habits of the cuckoo must be sought for in the operation of natural selection. The great difficulty is, however, to find the key of the situation. Why has the cuckoo developed in a certain direction and become such an exception to other birds? Many of the peculiarities which observers have taken for causes are without doubt effects acquired after the bird had already made progress in a certain direction. But what has been the starting point, and where are we to find the cause which first led to the development of the bird along such a peculiar line?

The proper point at which to begin an inquiry of this nature would seem to be that to which all the theorists are willing to return. There can be no doubt that the cuckoo, like all parasites, at one time lived a respectable existence. The bird must at some time or other have built a nest and reared its own young. There are many recorded observations of the reversion of the bird at the present day. to this long lost and aboriginal instinct of nidification. The cuckoo has been seen to sit on her own eggs on the ground, and she has been observed feeding her own young. It is even stated that she sometimes makes attempts at nest-building. Herr Adolph Müller has recently given an account of a case which he claims to have observed of a cuckoo hatching her own eggs. Comparing the cuckoo at the present day with other birds nesting under normal conditions, we find the parasitic habit associated with three remarkable characteristics. There are: (1) the undoubted gluttony of the bird and the peculiarity of its food; (2) the great preponderance of males; and (3) the extraordinary habit of the young cuckoo in the nest. Any theory of the origin of the cuckoo's habits through natural selection should be able not only to account for the parasitic instinct, but to explain in what way these peculiarities are associated with this instinct and with each other.

The only other bird in which the cuckoo's habits are known to be developed to a considerable extent is the American cow-bird. These birds exhibit in different degrees habits with regard to their eggs, varying from simple carelessness to the stage in which the parasitic habit is almost as well developed as in our own cuckoo. Some of the birds only show a disposition to lay their eggs carelessly about, occasionally dropping them in other birds' nests. In others the nest-building instinct has in great measure disappeared. The birds congregate together in flocks, and they often lay their eggs in heaps, so that only a small proportion are hatched, the parents assisting indiscriminately in the task of hatching the eggs and feeding the young. Lastly, in one species the cuckoo's habit is developed. A single egg is laid in the nests of other birds; the young stranger monopolizes the attention of the foster-parent; and though it is said not to eject its fellow-nestlings, like the cuckoo, these generally come by their death in consequence of its presence. The one noteworthy peculiarity which the cow-bird is said to have besides in common with the cuckoo is its gluttony. The bird is generally spoken of as possessing an insatiable appetite.

Returning now to our own cuckoo, there is one peculiarity of the young bird which seems very significant. It appears open to question whether the true meaning of the habit of ejecting its fellows from the nest has not been overlooked by observers. That the young cuckoo could have acquired this habit merely in order to be able to turn out the weak and small fellow-nestlings with which it is usually associated seems hard to conceive. These would beyond doubt be either starved or smothered in any case (as the companions of the American cow-bird appear to be) and it seems almost impossible that natural selection should develop so deep-seated a change merely to obtain so small an advantage. The alternative conclusion is that the cuckoo must have developed this peculiarity under conditions different from those now existing.

We may be able to realize to some extent what those conditions must have been, if we try to imagine for a moment what would happen if the cuckoo of the present day were to return to its aboriginal habits and endeavour to rear its own young. Two considerations immediately present themselves. A single pair would in the first place be quite unable to feed and rear an ordinary brood. In the second place, the young birds would not tolerate each other in the nest. There have been rare cases known where the cuckoo has deposited two eggs in the same nest, and one of the young birds has been known to eject the other after a prolonged struggle.

The cuckoo, whether from some change in environment, such as the disappearance of its natural food or through some other cause, is evidently at the present time a bird which finds great difficulty in feeding itself. Each bird is said to have its own feeding grounds, which it defends against all comers, and the early migration, and the significant fact that the caterpillars which other birds reject form the staple food, all point to the conclusion that the cuckoo obtains sufficient food only with difficulty. Now it is not difficult to conceive what the effect upon the young was when these conditions first arose and the cuckoo was still a normally nesting bird. The nesting period is the time when the demand for food is greatest, and the rivalry must immediately and in the first place have made itself felt among the young birds. The advantage must now have inevitably been with those birds which from generation to generation obtained the most food in the struggle which ever went on in the nest. This is where, in all probability, we must look for the origin in the young cuckoo of the habit of ejecting its fellows from the nest, and the development in the surviving birds, through the operation of natural selection, of the peculiar temperament which accompanies it.

If we are right so far, it is probable that we are now also in view of the explanation of the phenomenon of the great preponderance of males. It is a well-known fact that amongst most birds the males are always the stronger and more active in the nest. The advantage in such a struggle must always have been with the males, and the broods of which the greatest number survived were those of birds which produced the largest proportion of males. This selection may have continued after the cuckoo had acquired its parasitic habits. It would operate, it must be noticed, not simply by weeding out the females, but by selecting for survival the descendants of those cuckoos which produced a preponderance of males, and which would consequently transmit a similar tendency to their offspring. This tendency thus developed through an immense number of generations would inevitably become in course of time what we find it to be at the present day the normal habit of the bird.

The origin of the parasitic habit of the cuckoo is now less difficult to account for. We have here, in fact, only to follow in the main the explanation already suggested by Darwin, always remembering, however, that this habit is probably itself but an incident in a peculiar course of development rather than the cause of the development. The habit probably had its beginning in either one of two tendencies common among birds—viz., the inclination to steal from each other nest-building materials and the disposition to lay occasionally in each other's nests. Of the habit of stealing nest-building material we have a familiar example in the rooks, and it is widely distributed among birds. The Baltimore oriole, a near relative of the American cow-bird, is described as being very active in appropriating materials collected by other birds. Some of the cow-birds either build a nest of their own or seize one belonging to another bird, while in others, as already mentioned, the cuckoo's habit is developed. On the other hand we have many familiar examples of the habit of occasionally laying in other nests, especially among the gallinaceous birds.

It is not difficult in either case to imagine how the present habit of the cuckoo was developed, doubtless by easy stages. If the cuckoo of the present day finds great difficulty in feeding itself during its stay with us, it was obviously a great advantage for the voracious young bird to be entrusted to the care of foster-parents. The young birds developing from eggs which chanced to be deposited in the nests of other birds, stood a much better chance of survival, and this chance was further increased when but a single egg was laid in each nest. From a small beginning the habit would, in fact, be developed and perfected by the operation of natural selection alone.

There remains to be mentioned what is perhaps the most interesting example of gradual adaptation of means to an end which the habits of the cuckoo afford. Many strange and ingenious theories, a few examples of which have been given, have been propounded to account for the instinct which apparently leads the cuckoo to deposit her eggs in nests containing others resembling them in appearance. As a matter of fact, however, we have here only another beautiful example of appropriate results produced by natural selection. The great variation in the cuckoo's eggs has been already referred to. Anyone who has ever placed, as I have done, a cuckoo's egg of the largest type in a nest with the eggs of one of the smaller birds utilized, could not help being struck with the incongruity of the appearance. There would be little doubt in his mind that if the cuckoo herself deposited her eggs thus unsuitably, they must often not be hatched out. That this happens sometimes at present is not unlikely; that it happened more often in the past there can be little doubt. Mr. Nuttall relates significant instances of the sagacity of the American summer yellow bird in refusing to hatch the egg of the cow-bird placed among her own. The strange egg is sometimes broken, or being too large for ejectment, it is enclosed in the bottom of the nest and a new lining built over it, and the bird is said sometimes to enclose even her own eggs in this manner rather than hatch out that of the intruder. Some selection of this kind must undoubtedly have been going on in the case of the cuckoo's eggs for an immense period. The eggs which had most chance of being hatched out were always those most closely resembling the eggs of the foster-parent.

But now comes the most curious part. Natural

selection it may be said has acted thus far, but how comes it that a particular cuckoo lays a certain type of egg in a particular nest? The answer is very interesting. It has been noted by several observers that the same cuckoo always lays eggs of the same type, and recent observations also establish a strong probability that each cuckoo generally lays in the nest of the same species of bird. Now both these peculiarities would in all probability be hereditary. The cuckoo, in fact, deposits her egg in a suitable nest, not from any extraordinary or mysterious instinct, but because the descendant of a bird reared, for instance, in a skylark's, from an egg resembling those of the foster-parent, would herself probably lay in a skylark's nest, and produce an egg of similar appearance. We appear to have here an exceedingly interesting state of things. Natural selection has, as it were, developed in individuals of the cuckoo tribe the tendency to produce certain varying types of eggs, and at one and the same time has also developed the tendency to deposit these eggs in the nest of the suitable species of bird. The great variation in size and appearance in the cuckoo eggs, therefore, simply corresponds roughly to the variation among the eggs of the numerous species of foster-parents made use of by the bird.

If the facts have justified us in regarding the cuckoo as a bird which experiences great difficulty in obtaining sufficient food, we have found, therefore, in the operation of natural selection alone a sufficient explanation of the extraordinary series of habits and instincts which have rendered the bird remarkable from time immemorial. That the difficulty has been an increasing one from some distant time in the past the evidence seems to show. Whether it still continues to increase, and whether we must regard the bird as travelling slowly on the down grade towards extinction, it would be more difficult to say. Some of the facts may appear to point to this conclusion. The wide distribution of the bird, the extraordinary limits which it reaches in its migrations (it extends from South Africa through the tropics, and as far north as the pines go in Europe), the short period over which its visits extend, the nature of its food, and the well-known gluttony and rivalry for the feeding-grounds, are facts which, taken in connection with the parasitic habit, can at all events leave no doubt that the cuckoo of the present day maintains its position amid the competition of life only with extreme difficulty.

IX

THE HABITS AND INTELLIGENCE OF BEES

THE little busy bee has been a great favourite with the moralists and philosophers of this much-preached-at world. She and her works have been used to point so many morals to the intended disadvantage of the lord of creation, when his teachers take him to task in their sermons from the book of nature, that it is time some one undertook a serious examination of the claims of the little creature to be always posing as an example to the rest of the world. Not that it is to be expected that she would become less a subject of wonder and admiration, but rather because it would be interesting to be able to judge the exact amount of credit and respect to which she is entitled as an intelligent author of her own exemplary conduct.

There is no doubt at all events about the place of the bee in the insect tribe. In common with her cousins the ants, wasps, etc., she belongs to the order of *Hymenoptera*, ranking first in the insect series not only in the higher development of the cerebral ganglia, and general intelligence in habits and mode of living which this implies, but also in general completeness of form and structure. When bees are spoken of, the representative of the family most familiarly associated with the name is the ordinary honey-bee which has for countless generations lived, laboured, and died an ignominious death in the straw skeps of our rustic gardens. The common variety is often known as the German bee, its original home having been the woods and mountains of Central Europe. A successful rival of late for the notice of the intelligent apiarist is the Ligurian bee introduced from Italy, where in course of time, thanks to enforced separation from its relations north of the high ranges of mountains which hem in its native land, it developed those slight differences in structure and colour which now mark it as a separate variety. Both varieties were unknown in North America, until they were introduced from Europe; but they have thriven and multiplied enormously in their new home, especially in the Western States, where they are still known amongst the Indians as the white man's fly. The other bees known in this country are the humble-bees, of which there are several varieties; but, although very interesting in their behaviour and habits, as will be seen further on, these are but the bumpkins of the bee family, who are content to spend their rude lives in arcadian dulness, living from hand to mouth, with no capacity for the aspiring life and higher civilization of their more gifted relations.

I am not a bee-keeper in the proper sense of the word. In my opinion, that occupation, on a large scale at all events, should in this country be left entirely to those possessed of an unwavering faith in our variable climate. My bees are not required, as the British workman sometimes holds himself to be, to toil from early morning to night, that the fruit of so much labour may one day be thanklessly appropriated for the benefit of a greedy master. If they choose, they need trouble themselves little for the future ; for, if they have finished an unsuccessful season spent in rummaging the gardens of my neighbours around Clapham Common, the sweet stores of the nearest grocer are always liberally drawn upon for their benefit. One small colony is quite at home on a small stand in my room, having access to the outside through a little tube passing underneath my window-sill. The little creatures are, however, quite as anxious to get into the room as they are to go outside, for they probably think from experience that the world would be on the whole a very fine place to live in, if the good things thereof were within such easy reach as they usually find them when they are admitted from this side. Let me draw up the slide a little. There they are ; the little heads thrust expectantly forward, squeezing each other in the endeavour to force a passage underneath. One little amazon has pushed her way through ; and, as I want to introduce her to you, we will shut the door on the rest.

She is too much preoccupied rushing about in search of expected sweets to make her bow to the British public at the present moment. Look at her as she travels inquiringly round ; is she not a well-bred, intelligent-looking little creature ? Any one can judge for himself, without finding it necessary to take a slice of her little brain to look at through a microscope. Intelligent in every motion, clean cut, compact in form, with no gaudy patches of colour in questionable taste, but refined yet business-like in appearance—there is a general look about her which stamps her at once as belonging to the highest type of the insect race. We do not entertain a proper opinion of the importance of the little creature. In our dull way we are inclined to estimate her place in the world by the amount of sugar-water she and her tribe can contribute in the year, reserving a shrewd suspicion in the backgound that if the whole species were to be extinguished to-morrow it might unaccountably happen in these days of Yankee enterprise that the supply of honey in the market would be in no way diminished. But we greatly underrate the importance of our little friend. If the British nation were to be suddenly blotted out of the world, the even tenor of nature's ways would be very little disturbed; and, whatever the political world might do, the natural world would soon go on as smoothly and indifferently as if nothing had happened. But if our little friend the bee were suddenly to cease to exist, who shall describe the desolation and confusion which would invade the harmony of nature? How many shy flower-virgins, in plain and hillside, would droop and pine for her coming! How many noble, long-pedigreed families in wood and valley, finding life insupportable, would give up the struggle for existence, and become extinct! How would nature herself change her brightest hues and dress herself in sombre colours to mourn our little friend!

In these days of popular science it is hardly necessary to make more than passing reference to the part which the bee plays in nature. In the vegetable world it is a vital necessity that the fertilizing pollen from the stamens of certain flowers should be carried to the pistils of other flowers, and the mission of the bee is unconsciously to carry the precious dust from blossom to blossom in her search after the tempting drop of nectar with which the shy flowerets reward the winged bearer of their love-messages. A wonderful and fascinating chapter in natural history is that which treats of the relations existing between flowers and insects. Flowers may be divided into two classes, those fertilized through the action of the wind, and those in which fertilization is effected through the intervention of insects or a like agency. Darwin and others have shown what interesting stratagems flowers of the latter class resort to in order to secure the services of insects in this respect. Every little foible and weakness of the winged visitor is pandered to. What is commonly called a flower is indeed nothing more than a skilfully devised trap to attract the attention of insects, and thus ensure their services towards fertilization. Our little friend the bee is æsthetic in her tastes, and behold the varieties of flowers vie with each other to beguile her attention in the display of the most artistic blending of colours and beauty of design. She likes sweet scents, and the laboratory of nature is called upon to distil the choicest perfumes to humour her. But these are but an advertisement for the nectar which it is the principal object of the bee to obtain, and when she has alighted in search of it, it is only to find that the flowers have in many cases devised the most exquisite little mechanical arrangements whereby she is unconsciously compelled to effect the object towards the fulfilment of which they have indulged in such a lavish expenditure of beauty and sweetness. It is all effected in the simplest manner through the great law of natural selection, here seen in operation in its severe simplicity ; for the flowers of those plants which present the greatest facilities for fertilization get their seeds set, and so ensure the continuance of their species, while the unsuitable and unaccommodating kinds remain barren and are gradually weeded out. In a babel of tongues, and since first he found a voice, the poet has sung of the loves and sorrows of mankind, but nature still waits for him to interpret her heart ; if he ever learns to do so, there will be a new song in his mouth, for he will have a wonderful theme.

But nothing is perfect in this world, and I may, perhaps, be permitted a moment's digression here to refer to an instance on record of a wicked attempt to frustrate the design in all this adaptation of means to an end. My attention was first directed to the subject on the occasion of a letter which appeared in print some years ago referring to the export to New Zealand of two nests of our ordinary English humble-bees, in the hope that their descendants would come to the rescue of the colonists, who found that the red clover introduced from Europe would not set its seed and propagate its species in their country in the absence of the kindly help of the little attendants for whom it provides its honey. The writer expressed the hope that the humble-bees exported were not of a variety which he had observed had fallen into bad habits, in that the individuals, instead of obtaining the honey from the red clover in the manner intended by nature, had learnt to take unlawful possession of it by snipping a hole through the base of the tube containing it, without, of course, effecting the fertilization of the flower in the act. I have myself often since had my attention directed to this habit in these bees, and it appears to be well established that this propensity to subvert the purposes of nature is largely developed in humble-bees under certain circumstances, and not only in the case of the flowers of the red clover, but also those of the scarlet-runner and other plants. It appears, indeed, that our hive-bees also, if they are not actually guilty of the practice, do not scruple to take advantage of the easy access to the honey thus provided for them. Such practices, if they were to become the rule, would soon bring their own obvious punishment.

Like many of the disreputable shifts resorted to in trade, this habit is in all probability the result of fierce competition for the means of obtaining an honest livelihood—another example of the action and interaction of the various causes which silently produce change and progress in nature. The hive-bee, thanks to its habit of storing up food for winter use, as well as to the protection of man, is able to start work early in the year, and during the months of April, May, and June, it practically has the range of our fields and meadows all to itself. The colonies of humble-bees, however, store up no honey, and do not live through the winter, only a few of the young queens of last season surviving. In April and May the poor queen-mother has to seek out a retreat in which single-handed she proceeds to rear what only towards the beginning of July becomes a large family. Now when these issue forth to forage in the fields they find in many districts that, what with a host of competitors of their own kind—and the hive-bees, which are masters of the situation, having already turned the best part of the year to account —they can eke out but a very scanty subsistence, and so, like others in reduced circumstances, they take to the mostly illegal occupation of living by their wits. The humble-bee, no doubt, finds it saves time to obtain possession of the honey in the manner described, the stratagem in all probability being principally resorted to in order to forestall her rivals by obtaining first access to the honey stored in young flowers which have not yet opened of their own accord. This interfering with the purposes of nature is not to be commended, perhaps, but the poor humble-bees, for all that, deserve, in my opinion, considerable credit for the ingenuity thus displayed in seeking to hold their own under difficult circumstances in this hard world. Any one may convince himself of the keen competition which prevails amongst bees of all sorts towards the end of the season if he will take the trouble to observe our fields or hedgerows for a very short space at this time of year, or if he will count the number of times in an hour that a particular blossom is visited by a bee—or would be visited if it contained honey, as it is not necessary for a bee to alight on a flower to know that she must go away empty. Darwin has left it on record, after carefully watching certain flowers, that each one was visited by bees at least thirty times in a day, and it cannot be supposed that the little visitors in such circumstances find much to reward their industry. It has been also shown that they will often visit from twenty to twenty-five flowers in a minute. It is very interesting to note that on such occasions bees always keep to the same species of flower during each visit to the fields, a seemingly unimportant fact first recorded by Aristotle, which has acquired new significance since we have learned what is the true relation existing between the bees and the flowers they visit.

Is the bee entitled to the eulogies which have been lavished upon her for so long as a tribute to instincts which some naturalists have held to be little short of reason ? Entomologists of the present day seem to incline to the opinion that she is not. Despite the habits and wonderful social economy of bees, their acts upon analysis do not appear to be the result of such a highly developed intelligence as has been supposed.

For many generations naturalists have been loud in their praises of the architecture of the honeycomb, and they went into ecstasies when the mathematicians conclusively proved—after much disputing amongst themselves—that the bee in the structure of her hexagonal cell had solved the recondite problem of constructing her waxen storehouses with the maximum of strength and capacity combined with the minimum expenditure of material. Yet, however difficult it may be to believe it, it is now quite certain that the bee evinces no very extraordinary intelligence in producing the exquisite workmanship displayed in the honeycomb, with all its interesting arrangements of planes and angles. The first instinct of the bee was undoubtedly to construct a circular cell, and at present the work is always commenced by excavating a circular pit in the layer of wax from which the work proceeds. A moment's reflection will show that if all the cells were circular they would not fit closely together, and this would entail a great waste of space, as well as a large expenditure of wax in constructing a separate wall for each cell. Now, as the work of construction proceeds, both these undesirable contingencies are avoided in making the cell hexagonal, by simply straightening out, as it were, and eating away to a single thickness the original circular wall at the six points where it comes into contact with the walls of the surrounding cells.

If it were desirable to go into detail, it would be easy to show how easily and naturally this is accomplished in the manner in which bees work, and that without it being necessary to assume any extraordinary intelligence on the part of the little architects, who are guided by a few simple instincts, after the exercise of which the shape of the cell becomes a mathematical necessity.

Nevertheless, the honeycomb of the hive-bee is a wonderful instance of perfection in nature, and it has a place of its own in the story of evolution. Between it and the rude agglomeration of cells of the humble-bee there is a wide distance, and every step in the progress upwards has, no doubt, been taken through the operation of the law of natural selection.

The cells formed in the nest of the humble-bee arise in this way. The queen-mother commences by laying her eggs in a mass in a lump of matter composed of pollen and honey kneaded together, to form the food of the young grubs. When these are hatched out they burrow in the substance, and eventually spin their cocoons, and it is these cocoons, rudely fastened together with wax, which form the greater part of the irregular collection of cells found in the nests of humble-bees. When the young bees have emerged, the empty cocoons are used for the storage of honey, and it is only when storage room of this sort is not available, that the bees display their rude attempts at the art of cell-building in forming rough waxen cups to hold the surplus honey. These last are the only cells which the humble-bee actually builds, and in their structure it is not possible to trace even the rudiments of the wax-economizing art of the hive-bee.

In tracing the development of the highly finished work of the hive-bee from such a rude beginning as this, it is only necessary to remember how vitally important to bees is the art of economizing wax. It has been shown that the secretion of one pound of that costly material necessitates the consumption by the bees of from fifteen to twenty pounds of honey. It is easy to see, therefore, what an immense advantage it must have been to those colonies which long ago devised expedients for saving this precious material, and so were able to store up for winter use the large amount of honey which would otherwise have been consumed in its production. The advantage soon told in competition with other colonies, and so the progress was continued until the limit has been reached ; for at the present time, in the structure of the honeycomb, perfection has been attained, there being simply no room for further progress.

The question to what extent bees possess the power of communication with each other has engaged the attention of many observers. Experiments with bees, and also with ants, have thrown some light on this subject. It has been shown that the ants of a colony recognized each other even after a separation lasting fifteen months. The bees of one colony always recognize each other also, even after prolonged absence, and, although it has not yet been clearly established, there seems to be good reason to believe that they do so principally by the sense of smell, and not by a pass-word or signal, as has been supposed. There is no doubt that bees possess a very keen sense of smell, and they are perhaps guided by it in many ways which it is difficult for us to understand. They evince a very strong dislike to all bad odours, and show a general preference for those smells which are pleasing to us.

An amusing instance of the dislike of bees to bad smells came under my notice some years ago. At the time in question there was in my father's garden a plot of early potatoes, some distance in front of a spot where stood several hives. Early in the season the rooks commenced to help themselves to the potatoes, grubbing the young tubers out of the ground, and doing so much mischief that some had to be shot, and the dead body of one was impaled in the middle of the plot as a warning and example to the rest. Soon after this a most unaccountable fury took possession of the bees. No one dared to approach them, for they attacked and instantly put to flight every person or animal which ventured into the garden. This went on for some days, with most unpleasant results, and the bees were fast becoming a nuisance in the neighbourhood, when the mystery was accidentally explained. Some one happening to pass by the impaled rook in the evening discovered the cause and centre of all the mischief. Every exposed part of the poor bird's body, especially about the mouth and eyes, was literally bristling with the stings of hundreds of bees, which had sacrificed themselves in a vain and senseless revenge upon its offensive presence. As the little creatures always die from the injury caused by the loss of the sting, the destruction must have been considerable amongst the bees, who in this case fell victims to their own extreme sensitiveness of smell.

It is often assumed that bees possess the power of communicating to each other ideas of a complex nature; for instance, it has been stated that if a bee finds a store of honey, she will return with the news to her companions, who soon accompany her to share in the find. This is undoubtedly true of ants, but in their case the explanation is obvious, and observation and experiment leave no doubt that ants are guided principally by the sense of smell in following up the traces of a companion to the source from whence she has brought the food. This explanation, however, cannot be accepted in the case of bees, for it is not to be supposed that they could follow the track of a companion through the air by scent. It has not, however, been proved beyond doubt that a bee will lead her companions to a store of food in this way, though experiments point to the conclusion that bees can bring friends, though they have not the power of directing them, to treasures at a distance.

As we owe to the bees' taste in colours most of the artistic arrangement of tints in our bright-coloured flowers, experiments on the colour-sense in bees have attracted considerable attention. Experiments show that blue is essentially the bees' favourite colour; after which come, in order of preference, white, yellow, red, green, and orange.

That there are not so many blue flowers as might be expected is explained by the probability that all plants with blue flowers are descended from ancestors with green flowers, which, under the influence of what may be called bee-culture, have passed through stages of white, yellow, and generally red before becoming blue.

Although the vision of bees is very good in some respects, they show little intelligence in finding their way in certain circumstances. Sir J. Lubbock experimented with a bee which he put into a bell-glass, turning the closed end to the light, only to find that she generally buzzed about for a long time in a vain endeavour to get out at the closed end, while flies placed in the glass in the same way soon made their escape.

I have always found bees very stupid in this way. Last summer I placed a nest of humble-bees in a large glass vase, some fifteen inches in diameter, and nine in height. I kept the nest in my room, and, for several days after it was placed in position, the workers crowded towards the side next the light, making vain attempts all day long to get out, and this although the top was quite open, and the surface of the nest only a few inches below the rim of the vase. It was some time before I noticed any of the bees get out, other than by what could only have been accident, although I watched the nest for some hours daily. It could not be said that the change in position of their home had unduly confused the older bees, for those born while the nest was under observation showed the same want of intelligence, and up to the end of the season in the daytime a few bees were always at the side of the glass next the light, beating about in a vain endeavour to get out.

Bees do not seem to possess the feeling of affection or attachment ; even the respect for their queen savours of the coldest utilitarianism, and when through either accident or circumstances she ceases to be of use to the colony for the one purpose for which she is maintained, she is abandoned, or superseded, apparently without the slightest compunction or regret by her so-called subjects. Bees never seem to help each other in difficulty or distress, as is often done by ants. If you hold a bee captive by the leg, the others either take no notice of her struggles or do not attempt in any way to assist her. If you go further, and crush her to death, they quietly crowd around, and, in the most callous fashion, show their utter indifference by helping themselves to the sweet juices expressed from the body of their unfortunate companion. Yet if bees are fed regularly they often exhibit a kind of selfish friendliness somewhat akin to that displayed by the cats of the neighbourhood towards the cat's-meat man on his round. During several attempts which I have made to keep alive during the winter the queens of colonies of humble-bees, I have particularly noticed it in those bees.

I first tried keeping the bees in little wooden boxes, which I always opened at feeding time, allowing the occupants to walk about for a little before putting them back in their boxes. I was surprised to find after a little time how the bees expected to be fed when the boxes were opened, coming familiarly on to my hand in search of food, and making themselves quite at home. One royal princess I had who always made such intelligent attempts to escape on these occasions that I was obliged to discontinue the practice in her case, and I fed her instead through an air-hole in the lid of her box. I, however, continued to take out her box with the others, and after a short time I was much amused to find her generally thrusting her long flexible tongue through the hole in the lid as soon as she knew that feeding operations were going on, as if she would by this means remind me that I must not overlook her. This bee I used to believe had a brilliant future before her, and it was a matter of great regret to me when I was one day the unintentional agent of her destruction. In mild weather she used to be always on the watch for an opportunity to get out of her box, and one fine December morning when I lifted the lid she took a short flight across the room. In searching for her I accidentally crushed her on the carpet beneath my slipper, and so ended her brief career.

Sir J. Lubbock, after many experiments on the power of hearing in bees and ants, states that he never could satisfy himself that these insects heard any sounds which he could produce. In the case of bees it would be a great surprise to many to hear that they are absolutely incapable of hearing, and it must not be assumed that they are so because experiments have as yet yielded no satisfactory result. From time immemorial it has been the habit with rustic bee-keepers at the time of swarming to invoke the aid of noise to hasten the alighting of the bees. With some, it takes the form of drumming on a tin kettle, others beat candlesticks together, or even put their faith in the strains of a concertina or violin. Everyone has his own theory as to the object of this performance. One does it to overpower the hum of the swarm so that the individual bees may think they are left alone, and so make haste to alight. Another does it to keep the bees in the neighbourhood with the charms of the music; and a third hopes to drown the notes of the guides which may be ready to lead off the swarm to distant parts previously explored in search of an eligible spot to alight in. It is remarkable, however, that all agree in assuming that the bees hear and are acted upon by the noise produced.

Sir John Lubbock has recently tried a further series of interesting experiments to decide the question as to how far the power of hearing is developed in bees. To what extent music has power to charm the bee or guide her instincts may be judged from the result of an experiment of which he read an account at a meeting of the Linnean Society in November 1882.

Some honey was placed on a musical box on his lawn, and the box was kept going for a fortnight, during which time the bees regularly helped themselves to the honey. The box and honey were then removed out of sight into the house, and, although placed near an open window and only seven yards from the previous position, the bees failed to find the honey, although those brought to it in its new position afterwards found the way readily enough. He, however, declines to say that bees are incapable of hearing, and thinks it not impossible that insects may perceive higher notes than we can hear, and may even possess a sense or perhaps sensations of which we can form no idea; for although we have no special organs adapted to certain sensations, there is no reason why it should be the case with other animals, while the problematical organs possessed by some of the lower forms favour this suggestion. He is of opinion that the sounds which bees hear may be not the low loud sounds but the higher overtones at the verge of or beyond our range of hearing.

It is, however, remarkable that bees certainly do seem to hear on some occasions. The note with which the old queen threatens the royal brood as they come to maturity, and swarming time approaches, and so well known to apiarists under the name of "piping," can often be distinctly heard some distance from the hive, and is evidently intelligible to the young queens, for they respond in tones perfectly audible to the listener. Although bees will take no notice of a very loud noise even quite close to the hive, it is, however, remarkable that the slightest tap on the hive itself, or any of its attachments, or even a heavy tread some distance off, immediately disturbs them.

Despite the study and observation to which bees have been subjected, their habits and instincts are still a promising and most interesting subject of inquiry. The strange relation of the sexes has received more attention than perhaps any other subject connected with these little insects, both on account of the interest attaching to it, and also because of its bearing upon other questions. The subject is, however, still full of difficulty, and the more it is investigated the more the interest attaching to it seems to grow.

In a colony of bees there are the drones (males), the queen (female), and the workers (neuters). It has long been known that the neuters are merely imperfect females, and the bees possess the wonderful instinct which leads them, in the event of the loss of their queen, to take a young worker grub or egg, and, by special feeding and the enlargement of its cell, to rear from it a new queen. It has been proved that parthenogenesis always prevails in the production of the male bee, the egg which produces a drone being always unimpregnated even when laid by an impregnated queen. A virgin queen will also lay eggs abundantly, and it has been conclusively proved that these eggs will come to maturity, and that they will invariably produce drones. Now, the bees always build a certain quantity of what is called drone-comb, in which the cells are larger than ordinary, and it is in these cells, and in these only, that the queen lays the eggs which produce drones. A knowledge of this circumstance first led to the assumption that the sex of the young bee was determined simply by the size of its cell, but this theory was soon abandoned, as it is settled beyond doubt that the sex of the egg is determined at the very moment at which it is laid. The theorists were then driven back on an ingenious explanation as to the mechanical effect of the shape of the cell upon the queen in the act of depositing the egg. This view has, however, also been rendered untenable by the result of experiments which place it beyond question that the sex of the eggs is altogether independent of the shape or size of the cells in which they are laid; for, with no drone-comb, the queen will sometimes lay drone-eggs in worker cells, from which eggs drones will be produced, and she will also, if necessary, though with great reluctance, lay worker-eggs in drone-cells. It would thus appear that we must concede to the queen bee the surprising instinct or intelligence which enables her to lay at will a drone-egg or a worker-egg, for in the hive she often passes immediately from the worker to the drone cells or *vice versâ*, depositing an egg at the bottom of each which always produces a bee of the sex intended. This instinct is rendered more wonderful when it is remembered that the number of drones produced in a hive is always regulated by the wants of the colony. The questions suggested by the manner of the production of the worker-bee are also highly interesting. It has been mentioned that the bees, when they require a queen, will take a worker-egg or grub and by especial feeding rear from it an ordinary queen bee. It has generally been stated that the young queen is in such cases fed with richer food known as royal food, but it seems by no means unlikely that we shall soon learn that this is slightly incorrect, and that the queen grub is in such cases simply fed with as much food as it requires. This would mean that the queen state is that to which all the worker-grubs would develop in normal circumstances, and that the bees deliberately and for social reasons prevent this natural development by a *régime* of low diet. One who has made a special study of bees gives it as the result of his observations that the bees feed the worker-grubs sparingly, as if fearing an excessive development—a truly wonderful instinct which has enabled the bees to solve one of the most difficult of social problems. In the construction of the honeycomb the bees anticipated the mathematicians: have they not here again anticipated the philosophers?

THE HABITS OF FROGS

THE mild days of March witness in many of the more secluded parts of the country a sight which is not to be quite matched by anything in nature. Frogs pass the winter in a state of torpidity. They bury themselves deeply out of sight in moist banks or beneath the roots of trees, or, best of all, in the spaces between large stones loosely piled together. Great numbers are sometimes dug out of such retreats in winter, and they generally lie packed and flattened together almost in a solid mass ; probably, like hibernating bees, in order that the temperature may be kept a degree or two above a minimum-point at which it soon becomes fatal. Frogs when they retire for the winter take leave of food, for they eat nothing for four or five months. Yet the remarkable fact is that when they emerge from their long sleep in the month of March the females are ready to spawn. Both sexes almost immediately make for water, and it is one of the most remarkable sights to come across a meeting-place of a certain established kind. Everyone of course has seen occasional frogs seeking water in early spring. This is not what is meant. Hidden in woods and marshes there are pieces of water in the country which at this season have been trysting-places of frogs for hundreds, and it may be for thousands, of generations. From far and near the frogs in early spring move towards these spots by unerring instinct. The surface of the water may sometimes be seen broken and rippled, as if a shoal of mackerel were underneath. The writer recently saw such a meeting-place where the frogs thus collected must have numbered thousands, and the croaking, splashing, and gurgling of the creatures blended together into a curious body of sound which was audible a long distance away.

Strange as it may seem, there is a certain fascination in the frog as a tame creature. There is probably no other animal which has been so much studied and experimented on by science as the common frog. Yet it is surprising how little we know about the personal side of him. Taken young in the spawn stage, he develops into a tadpole through a series of uncanny metamorphoses which, while they certainly suggest as seen under the microscope strange scenes and climates in the past history of the world, do not lend themselves much to a study of his elusive personality. It is only when the creature absorbs his tadpole's tail (it does not drop off) and emerges from the water a little frog about the size of a sixpence that his true personality can be said to begin. The writer has kept many little frogs from this stage upwards. They soon begin to take an interest in their surroundings and to show intelligence in their habits. They will readily feed on small grubs, worms, and insects. The frog is peculiar in one respect about his food.

It must not only be alive, it must be stirring, otherwise he will not touch it. The tongue is shot out at a lightning pace, and retracted with equal speed, taking the prey with it. The frog will get to know persons, and there have been instances recorded in which they have answered to their names and would come to be fed when called. Frogs certainly have very highly developed gifts of instinct or intelligence of their own kind. When necessary they are, for instance, able to make straight for water even from a very long distance. The large bull-frogs of America are almost as intelligent as rabbits. The little French climbing tree-frogs have also a very keen sense of locality. The writer once, after much effort, captured one of these, and, taking it home, placed it, apparently securely fastened up, in a large room. During the night it got loose, climbed the wall, and finding the only opening in the room, escaped through an inch of space which had been left over the top of a window-sash. Frogs will show no mark of satisfaction when pleased, but our common frog can express feelings of fear or terror in a most striking manner. If chased, especially by an animal which it fears, it will sometimes, if it thinks escape to be impossible, give utterance to a squeal which is almost human in its pitch and intensity. All frogs have a keen appreciation of coming changes of weather, and they will adjust their habits accordingly.

It is not easy to say, in the case of the frog, how much is due to intelligence and how much to instinct or even to reflex action. It is one of the commonplaces of science what a frog will do even after his brain is removed. If care is taken to keep them alive after the operation, brainless frogs will catch flies and even proceed to bury themselves in the earth at the beginning of winter. A decapitated frog will jump naturally; if it be placed on a table and if irritation be applied, it will scratch at the place. If the irritation be continued the itching apparently at last becomes intolerable and it will make a most natural dive for the floor. The frog in this respect is not peculiar. Dr. W. H. Thomson tells an amusing story of how when fishing out west in America the bait was taken by a mud turtle which swallowed the hook. Being unable to get the hook free the angler hung the turtle over a branch and sawed the head off with a pocket-knife. Down dropped the turtle's headless body, when to Dr. Thomson's astonishment it straightway walked some two yards right into the water and dived into the deep pool just as if the creature's brain had still been directing it.

The frog very nearly resembles the toad in many of its habits and instincts. In one respect, however, it seems to differ considerably, as any one may readily prove. A toad will survive a long period, certainly more than a year, buried without food and almost without air. The frog soon dies in similar circumstances. It used to be a matter of mystery as to how frogs after hibernation came forth in the spring well nourished and ready to spawn. The explanation is now fairly evident. In the autumn, frogs consume great quantities of food, earthworms and snails being their principal prey. Frogs and all their kind possess internal organs which are technically known as fat-bodies. These become richly stored with globules of fat and oil. Their function is now known to be that of reservoirs. In the spring they enable the spawn to be ripened by drawing on the food material which they contain. Frogs do not feed during the excitement of the spawning season in March, and if it were not for this effective resource their bodies would certainly sink exhausted during the exceptional strain of this period.

XI

SEA TROUT

AS the long summer daylight has slowly waned with the past month or two the angler has assiduously courted that sporting trout which, having moved in deep waters, has at this season returned to his inland haunts with the mystery of the sea upon him. On the larger waters of the north the angler has gone out morning after morning in the early light, with only his trusty boat-man. Rowing far and taking note of wind and weather he has, if possible, set himself a silent drift for miles with his back to the wind while he has whipped the grey waters hour by hour with the gaudy lures that are sufficient to draw *S. trutta* from the depths. The salmon trout or, as he is more generally known in the north, the sea trout, is the most esteemed of all our fishes after the lordly salmon. Sport he gives in plenty when he is hooked; and excitement and trial above all his kind in that lightning moment ere he rejects, as he often does, the lure which brought him from below. He is born in fresh water like the salmon, and like the salmon that is to be he passed the first epoch of his life-history a little untravelled trout knowing only the shallow pebbly bottom of the upland streamlet in which he found himself. He had to all appearance no knowledge of the sea even of that kind which creatures are said to have in dreams, and he would choke and die if placed in salt water. But the time came when he lived in another world. The lust of life came upon him and he travelled. Now when the angler meets him he is back from the unknown with the grey and silver livery of the sea upon him, a salmon, or all but a salmon, in appearance. Large he has grown, plump and strong and full of the fight of life. Very different must have been the fare upon which he lived in the sea from that which his mountain eddy provided, yet here he is now returning to meet his mate in the mountain burn and with all the old life being slowly reimposed as it were upon him.

It is one of the most interesting of life-studies how it can all possibly happen. Here, for instance, when the skilful boatman drifts before the wind, there is a long reach of particular bottom. Elsewhere the waters of the loch may be deep, but here the point is that they range from a few fathoms to a couple of feet, and that the ground is pebbly or stony. It is the kind of bottom which produces feeding which reminds the returning trout of his early haunts. He thinks again of the larvæ of water-flies and water-beetles. He rises after the winged and perfect creatures almost as soon as he has left the sea. Swish! he rushes upwards like a streak of silver to your own teal and yellow monstrosity, "coming short," as the sea trout so often does. In a little time more these same fish will be back in mountain streams, where there is water scarcely deep enough to cover them, fish of two pounds, of five pounds, of ten pounds, meeting their mates, scratching in the gravel, depositing their spawn. And then returning once more, spent and exhausted, to the sea.

What is the real explanation of this instinct which drives the sea trout, like the salmon, thus to return from their sea-going life to spawn, not merely in fresh water and in running streams, but in particular places and in quite shallow water, often of a few inches? It is interesting to note in most hatcheries how extremely significant are the conditions under which the eggs of salmon trout and salmon have to be hatched out. The eel returns from inland waters to spawn in the depths of the ocean. Probably considerable pressure is necessary to its eggs, as it is to the eggs of many sea fishes, before they will hatch out. In the case of the migratory trout the fact which probably determines the curious life-history of the fish is the very interesting, although withal very simple one, connected with the eggs. The eggs of the trout will not hatch out in any but the shallowest water. The least pressure is quite fatal to them. This is a fact well known at all artificial hatcheries, and all arrangements have to be made accordingly. It was no doubt around this little tuft of circumstance that the strange and eventful history of the salmon family slowly evolved itself. When the trout took to migrating to the sea or the river estuaries at first there probably was no universal instinct to return to their old haunts for spawning purposes. For ages there must have been, as indeed there may be now, salmon trout tending to shed their spawn where they lived, in the rivers, in the estuaries, in the seas. But in the nature of things no grain of it has ever come to maturity. Only the fish which happened to return to spawn to the shallow upland waters have ever left any descendants. Hence the rigid process of long-continued natural selection which evolved the present habits of the salmon and which every year drives all the existing members of the salmon family to their spawning-grounds. Hence the imperativeness of the instinct to which they are subject. They are, as it were, the survivors of a vast army which must have become extinct through lack of the instinct now developed in them.

Salmon trout are persistent travellers within a certain limited area. A fish of a few pounds weight which rises to the fly of the angler has in all probability journeyed through the same waters before. The salmon moves far in its migrations and even goes far to sea, but the sea trout has a more local range. For instance, a considerable proportion of marked fish have been known to return to the same small stream and to exactly the same locality the year following that in which they were marked. It is one of the most interesting features of the migration of salmon trout that large fish of several pounds weight will ascend quite narrow rivulets for spawning purposes. If one may judge, indeed, by facts obtained from artificial experiences, the conditions which the fish seek for successful spawning are not easy to find. Not only must the water be extremely shallow and running, but the eggs cannot be hatched out in strong light, nor will they come to maturity except the water be kept at a low temperature. The salmon trout, therefore, seem naturally to deposit their eggs in dark or shaded situations and between banks which run east and west rather than north and south.

We may see also why it is that the salmon trout which the angler meets like the salmon in certain waters in certain seasons has reached the place where he is found at that particular time. The eggs of his mate will hatch out only in water at a temperature not high above freezing-point. All the flittings of the fish must therefore be arranged so that he may arrive at the spawning-ground in the colder months of the year, according to locality. Others may suit their journey to more convenient seasons; for him it is expedient that his flight should be in winter. It is probably for a similar reason that we find many of the movements of salmon so puzzling. It has often been pointed out that it cannot be simply in obedience to the sexual instinct that the salmon ascends the rivers, for many of the fish have little development of roe when they leave the sea. But when the object is kept in mind, it is all quite clear. Some of the fish ascend rivers to spawn which are but a few miles long, while others have to travel to their destination for hundreds of miles. In all cases the spawning time of the salmon, like the sea trout, is in the same months of the winter, so that the journeys which it undertakes must be arranged accordingly.

XII

THE INSTINCT OF ANIMALS

I WAS standing in the small zoological garden in Pretoria, established by the late President Kruger for the instruction and amusement of his people. I had been in the place some time, and was watching half a dozen monkeys which were chained to trees inside a wire fence. One of these had attracted my attention. This was at first merely because of his behaviour to a certain class of visitors. It was soon after the South African war, and some of those who were strolling through the place were soldiers who had taken part in it. This particular monkey, whether by training or otherwise, always flew into a great rage at the sight of the British uniform, chattering furiously, and grinning at the worn khaki-clad men who from time to time approached the rails to look at him. The visitors had been throwing various kinds of food to the monkeys, and some of it had fallen beyond the limit of the chains and lay on the ground out of reach. The other monkeys kept each near the foot of his respective tree, but this one remained on the ground near the railings with an air of preoccupation which was noticeable. Now and then visitors approached who carried sticks, and presently one of these was dangled loosely outside the rails. In an instant the monkey had snatched it, and had fled chattering with it to the foot of his tree. I expected to see him examine it curiously, and perhaps break it in pieces, as is the way of monkeys. But no; he had a more deliberate aim. Advancing on all fours to the limit of his chain, and with the stick stretched out in his hand, he proceeded, amid the greatest excitement among the other monkeys, to rake in, one by one, the titbits which had accumulated, hitherto beyond reach. Although there was nothing new to me in the act, for I had previously lived with and studied monkeys at close quarters, I shall not forget the effect for the moment on my mind, and on the minds of some of the spectators as I saw it reproduced in their faces. Had not the zoologists been right in placing the monkey among the primates? Here was something more than mere animal instinct. Was not this an example of mind conquering the dull tyranny of things as they are, and the first tool-using animal emerging beneath our eyes?

It has been my experience to be able to study animal instincts and animal intelligence, both in the lower and higher animals, in many conditions, for a period now extending over more than twenty years. Deep and lasting, on the whole, has been the impression left as to the results of animal instincts. Nevertheless, it yields place to a deeper feeling as to the character of the enormous interval which separates the highest example of animal instinct from even such a simple act of intelligence as that recorded above. The most permanent result of my own studies in animal capacities has been a gradually increasing conviction as to the as yet unimagined significance of mind in the further evolution of the universe. But I think that a first step toward a truer appreciation of the almost inconceivable potentialities of mind in the future is a clearer perception of the difference which marks off its higher manifestations from even the most remarkable examples of animal instinct.

On the table before me is a little red-covered box connected underneath the window-frame with the outer air, and corresponding in some respects to that used by bee-keepers when they wish to start a new colony. It contains a single full-sized comb filled with bee-brood in all stages of development, which was transferred four days ago from an ordinary bar-framed hive. The excitement now visible among the bees as I let down the shutter is intense. In one spot they have broken down the worker-cells and are building up a large structure, at the bottom of which one catches occasional glimpses of a white grub that an eager crowd of bees, jostling one another in their excitement, are endeavouring to feed. It is the new queen that is to be. This action of the bees in thus, on the loss of the mother of the hive, selecting a worker-grub and rearing from it a new queen, has often been described. There is no example of animal capacity in nature which is more striking or which has been oftener remarked upon as displaying reasoning power. It exhibits an apparent intelligence and foreknowledge which call forth the admiration of all observers. Nevertheless, it has to be admitted that the same bees in other circumstances show no particular gifts of intelligence. Bees when short of food readily steal honey from other hives. But in a great number of experiments which I have tried in which the queenless bees have been left without either egg or larva from which to begin a new queen, I have never had a case in which they have attempted to avoid extinction by obtaining a larva or egg from another hive, which they might easily do. Bees have a wonderful instinct for finding their way. Yet out of their usual habits they readily lose it. If one among her companions is killed, the others exhibit neither fear, nor resentment, nor interest. If one is provoked to use her sting, she makes no intelligent attempt to withdraw it, as she sometimes might do, but walks away, stupidly dragging out her entrails and causing her own death by the act.

The instincts by which migratory birds find their way, year after year, for thousands of miles, over wastes of sea and land, appear to us little short of marvellous, and seem often to indicate in like manner a high order of intelligence. Yet it by no means follows that we are witnessing in these cases also any more than a mechanical or unreflecting response of the organism to its environment. Why the powers appear to us so wonderful is that we do not always know the exact nature of the stimulus, and possibly do not ourselves possess, or possess only in a very rudimentary form, the senses which are concerned in responding to it. The attunement of an organism in this manner to the calls of its environment, through senses which are beyond us, but senses which compel it to do mechanically what in the higher animals is done by intelligence, is, nevertheless, one of the most wonderful products of natural selection.

Some years ago it was my good fortune to rear from the beginning a specimen of the young of the common cuckoo. The habits of this migratory bird, which is a spring visitor in Europe, though it is not found on the North American continent, are well known by repute. The female lays her eggs in the nests of small birds, and the young cuckoo, when only a few days old, and while it is yet blind and almost naked, ejects from the nest, with a purposiveness which is almost uncanny to watch, its fellow-nestlings, and receives thereafter the sole care of its foster-parents. As my young cuckoo became full-grown, the degree of complexity and perfection obtained by nature in mechanically attuning this bird to the wants of its curious migratory life was extraordinary to witness, and made an unusual impression on my mind. The cuckoo, it may be mentioned, travels, in its annual migrations, enormous distances over land and sea, sometimes from the extreme north of Europe, across the equator, into the Southern Hemisphere. In this case there is no room for thinking that the young birds find their way as the result of any teaching from the older birds, for these leave many weeks later than the older birds, and so travel apart.

As the season waned, and the time for the migration of my young cuckoo approached and passed, its behaviour grew interesting. The bird always became very restless in the evening. Being much attached to me, it generally settled at last, so as to be near me, on the stationery case on the table on which I was writing, in the dim light thrown by the upper surface of the green shade of the reading-lamp by which I worked. Here, as the hours wore on, the same thing happened every night. After a short interval the muscles of the wings began to quiver, this action being to all appearance involuntary. The movement gradually increased, the bird otherwise remaining quite still, until it grew to a noiseless but rapid fanning motion of the kind that one sees in a moth when drying its wings on emerging from the chrysalis. This movement still tended to grow both in degree and intensity, and it generally lasted as long as I sat up during the night. In the early stages of this mood the bird responded when I spoke to it ; but in time it ceased to do this, and became lost in a kind of trance, with eyes open and wings ceaselessly moving. Brain, muscles, nervous system, and will, all seemed inhibited by the stimulus that excited it. The bird became, as it were, locked in the passion of that sense by which the movement of flying was thus simulated. It was one of the strangest sights I have ever witnessed—this young migratory creature of the air which had never been out of my house, and which had never known any of its kind, sitting beside me in the gloom of our Northern winter, and in the dim lamplight, and by a kind of inherited imagination, which was yet not imagination in our sense, flying thus through the night, league-long, over lands and oceans it had never seen.

There was, I think, no question of the exercise of intelligence in this case. What had rather to be noticed was the entire physical system of the bird thus hereditarily attuned, and in an inconceivable degree of perfection, to react to stimuli related to the necessities of its migratory habit of life. In instances like this the stimuli to which the organism responds, as a wound-up spring responds to the touch which sets it loose, are mostly beyond our senses. We may form some conception of them as being probably similar in character to those which enable the sexes of various animals to distinguish each other, and often, in certain forms to find each other even when immense distances apart. These stimuli are, no doubt, of various kinds. Sometimes, as when one sees birds migrating south against a head wind, they may be related to a highly sublimated sense of smell, or something akin to it. It may be mentioned in this connection how even a slight peculiar odour often serves in our own case to evoke immediately and powerfully ideas and emotions associated with it ; and it would seem not improbable that we have in such a fact suggestion of the rudimentary survival in man of a faculty which has played a great part in the evolution of life on a lower scale.

One of the senses in birds and other animals giving rise to remarkable results, is the sense of direction. Darwin relates an incident of a horse which he had sent by railway from his home in Kent, over a hundred miles westward to the Isle of Wight. "On the first day that I rode eastward," he continues, "my horse was very unwilling to return towards his stable, and he several times turned round. This led me to make repeated trials, and every time that I slackened the reins he turned sharply round, and began to trot to the eastward by a little north, which was nearly in the direction of his home in Kent." Darwin concluded that the horse knew, even at so great distance, the direction in which his old home lay. Similar facts will be familiar to most persons who have had experience with horses. They have been related to me in great number by cattlemen in the Western States of America, and by Australians who have lived in the bush. A similar faculty is highly developed in dogs, cats, and other domestic animals. It has been proved not to be due to any conscious noting of landmarks, for animals have found their way back over immense distances, even when they have been sent on the outward journey in closed boxes.

It is interesting to note that this faculty of judging direction seems to bear no relation to the place of the animal in the general scale of intelligence. It is possessed to a considerable degree by dogs and cats; but it is possessed in a very high degree by seals, who find their way unerringly back every year to their rookeries from enormous distances to which they disperse in the open sea. It reaches an extremely high degree of perfection in migratory birds not otherwise noted for intelligence. Even animals low in the scale, like fish, find their way regularly for great distances to their spawning-grounds. A case is related of a snake, carried in a closed carriage from Madras to Pondicherry, a distance of 100 miles, which found its way back.

Possibly we have in man in this case also a suggestion of the survival in rudimentary form of a faculty far more highly developed in lower animals. Those used to the open life of the West have told me that many of those bred to it come to carry with them, even when out of sight of all landmarks, an overpowering instinct of direction. In the case of men this sense appears to work by a process of sub-conscious dead-reckoning in the mind; but in the case of animals it is doubtless often supplemented and directed by stimuli which do not reach our senses. The character or direction of the light, changes of temperature, nature of the wind, odours, emanations, or radiations which mean nothing to us, or which do not reach our senses at all, may profoundly influence animals whose safety or welfare for innumerable generations in the past has depended on correctly interpreting in action the message which they convey.

That the clue or explanation in all these cases of instinct is a comparatively simple one, if we only knew it, is, I think, highly probable. Recently, in South Africa, in discussing the scouting during the late war with one who had acted as a scout, I challenged a test. We were taken out on a dark, cloudy, and still night, turned round many times, and at last asked to point in the direction of the place whence we had started. My companion failed; I succeeded without hesitation, and I seemed to him for the time being to be endowed with a special and unaccountable sense of direction. What I had done, however, was simply to experiment with a trick known to poachers and sportsmen in England. I had wetted my fingers, and, holding them up, was enabled to distinguish the direction of the very slight air current. My inexplicable gift was, in short, due to no more than the simple device which had rendered me, for a moment, extra sensitive to the direction of the wind.

However wonderful and however inexplicable animal instincts like the foregoing may be, they are not, I think, usually accompanied by the exercise of any high degree of intelligence. They represent rather the mechanism of mind in an early stage in the evolution of life. The instincts in question are always born with the animals ; but what is inherited is not, as is sometimes imagined, knowledge or ideas ; it is simply the physical organization, common to a whole species, adjusted, often with exquisite perfection, to respond more or less mechanically to stimuli related to the average welfare of the animal. Why the instinct often appears wonderful to us is that we do not possess the same organization, and that the stimuli to which it responds are therefore often beyond the reach of our own senses.

A more noteworthy class of instincts than these belong to a higher class. Most students of wild nature in northern Europe, Asia, or North America will have made the acquaintance of the wild duck from which our common domestic duck is descended. If this shy bird is surprised in the spring in sedge or reeds with her young, she possesses a peculiar habit which is interesting to watch, and about which Darwin and Romanes held some difference of opinion. I have many times witnessed the habit myself. If one comes on the mother bird in wading through the sedge, she first attempts to escape through the cover without attracting notice. As soon as the eye catches her, she is seen to be swimming rapidly in front, followed by the brood of ducklings, the latter packed together so closely that they seem to move through the water behind her like a solid bank of dark-brown fur. The moment the mother duck perceives that she is seen, she springs clear out of the water. Not, however, to fly away ; for, as you see, she falls back, painfully flapping a broken wing, and beating heavily about in her attempts to escape. The eye always momentarily loses sight of the young birds in following these movements, and the ducklings invariably have disappeared when it seeks them again. It is next to impossible to find them afterward, although they are never far off. As to the final scene, I quote my own words as I have used them elsewhere to describe it: "As you continue to move, you notice that the unusual exertion is having a wonderfully curative effect on the broken wing of the mother. She is already taking short flights with it, still occasionally flopping back heavily into the water. As you look she sits up and flaps both wings airily enough. Now she springs into the air, and wheeling several times nimbly overhead, actually takes her departure altogether, with a series of wild derisive quacks as a parting salute. You feel somehow as if you had not got the best of the encounter, and that you have been treated throughout as a creature of inferior intelligence."

Darwin, in explanation of this instinct in the wild duck, thought that it was impossible to conceive the mother bird as consciously imitating the actions of a wounded duck, for she, in the vast majority of cases, could never have seen such. The original groundwork of the habit he considered to have been such action as one sees in the common hen, which, when her chickens are approached by a stranger, rushes excitedly about with ruffled plumage and extended wings. Natural selection, he considered, had accumulated in the wild duck those variations of this habit in which the actions of a wounded bird were mimicked, until in time a fairly good imitation resulted. The instinct was, in short, an example of those more or less perfect but unconscious and mechanical adjustments to the conditions of existence that we have been already discussing. Romanes, however, went further than this ; he agreed in the main with Darwin's view as to the origin of the instinct, but he held that direct intelligence in misleading an enemy now played a considerable part in the exercise of this peculiar habit on the part of the mother wild duck.

I am inclined to think that this view of Romanes is correct, and that there may be conscious, deliberate, and individual intention on the part of the mother to deceive, and so to adjust her actions as to cover the retreat of her young. Experiments which I have made since this article was begun with wild ducks which I have tamed, have, however, led me to think that there is a physical cause for the action, and that the conscious purpose may be only secondary. I have purposely chosen this example because it well illustrates the way in which an instinct of the lowest class merges into a higher order of intelligence. But I have also noticed it because it gives us a probable clue, to be referred to presently, as to the conditions favourable to the display of individual intelligence in animals in a large class of examples that attract attention.

As we rise in the scale of animal intelligence, there are two leading facts which have to be noticed. In the first place, it has to be observed that the organized mechanical response to stimuli which has been so far described, and which constitutes instinct in its lowest form, becomes replaced by something higher. We begin to have conscious intelligence in the individual initiating and directing action in such circumstances as may arise, and doing this with a growing perception of the relations between cause and effect. In the second place, it has to be remarked that zoological affinity does not indicate the line of this upward advance. The rat and the beaver, for instance, which furnish two of the most notable examples of animal intelligence, belong to a group comparatively low in the scale. The positions of the horse, the dog, the parrot, and even that of the elephant and the monkey, are similarly not clearly suggested by their structural affinities.

The conditions just referred to as favourable to the display of individual intelligence in animals are, I think, worthy of closer attention than they have hitherto received. Civilized man thinks so readily, and so easily, that we do not realize what a special and concentrated effort the mere rudiments of thought must imply in an animal. The expression of unutterable weariness which overspreads the faces of some savages when they are asked questions requiring a little mental effort has often been recorded by observers. How much more should we expect thought to require a supreme effort in an animal. In the valuable series of experiments recently conducted it is interesting to notice the efforts made to fix the attention of the animals, and how difficult it often was to retain it.

It may be observed that in most of the striking instances of individual intelligence that are from time to time recorded, there is a condition that is usually present. The circumstances are nearly always those in which some overpowering cause tends to concentrate the animal's mind on one subject. Among ourselves it is well known how such a condition stimulates thought. Every public speaker who has held a large audience knows how it tends to promote the flow of ideas. In animals, in those cases of exceptional individual intelligence referred to, it may be noticed how often this condition prevails. Cats, birds, and almost all animals, may be noticed to exhibit, as in the case of the wild duck just mentioned, greatly quickened intelligence in their actions in supreme crises where the safety of their young is concerned. Similarly, in cases of extreme individual danger, or of great desire, we often get marked instances of animal intelligence. Where both these conditions are combined, as in the attitude of wolves, jackals, foxes, and other animals to traps, we get those surprising instances of intelligence which are recounted, and often received with incredulity, though they are probably in the great majority of cases quite true. The experiences of professional trappers in northern Michigan give instances of foxes acting so intelligently in regard to the mechanism of the ordinary spring-trap that they consistently burrowed beneath the jaws so as to push down the pan from beneath, and thus spring the trap with safety.

Similarly, in cases of concentration of attention under the influence of strong emotional excitement, as when the animal is wounded, there is often evidence of great stimulation of intelligence. Menault relates the case of an eagle caught in a trap and afterwards undergoing a surgical operation. " Though his head was left loose, he made no attempt to interfere with the agonizing extraction of the splinters, or to disturb the arrangements of the annoying bandages. He seemed really to understand the nature of the services rendered, and that they were for his good." Most persons will have had, some time or other, similar experiences with cats or dogs, or in relieving animals caught in traps. Romanes quotes a number of instances of elephants intelligently submitting to surgical operations, bracing themselves against pain and, in the words of the operators, behaving as if they understood the object of the acts. Monkeys when wounded are almost human in their behaviour.

The impression left on the mind by long and close study of animal instinct and intelligence is apt to be different from that which is popularly conceived.

When all due consideration is given to the powers possessed by the higher animals, one is, I think, impressed most in the end by the enormous interval of progress beyond this which the human mind so evidently represents. When it is considered how naturally it comes to man to use tools, it seems matter for surprise not that we should occasionally see this faculty in animals, but rather that we should so rarely, even in the higher animals, see intelligence rise to this level. When we observe an elephant prepare a branch to switch off flies, or see a monkey use a stick to rake in nuts, as in the example quoted at the beginning of this article, we are much impressed. Yet how far off, after all, do these efforts leave the animal mind! The monkey, although he rises to this level, will sit and warm himself at a fire without ever grasping the relationship between the fire and the fuel which feeds it. He will be the intelligent companion of man, and yet all his life never reach the communionship of the most elementary forms of speech. An intelligent bird like the parrot, on the other hand, will articulate human language perfectly, and yet with no mind behind to furnish the link between the spoken words and the ideas they represent.

The relationship between cause and effect appears to the human mind so self-evident that in certain systems of philosophy it is regarded as the most elementary and fundamental of all knowledge. Yet there is certainly no conception of it in the minds of most animals. The common domestic fowl will suffer the utmost inconvenience in tossing loose pieces of green food over her head in the endeavour to break off morsels small enough to swallow. After a time she learns by experience that pieces under her, upon which she happens to be standing, are conveniently fixed, and she will look for them there. Yet for two seasons in which I had a group of Buff Orpingtons under close and almost daily observation in relation to this fact, I never once, during the hundreds of times I witnessed the act, could be sure that I saw any of them connect the cause with the effect, and consciously grasp with her claw, or hold down in position, a piece of food. Green food did not grow in detached pieces in nature, and the automaton, adjusted to nature, contained, therefore, no response. Even where animals perform such acts with apparent intelligence, we never can be sure how far the result is due to inherited reflex action. We ourselves blink at a blow which threatens the eye, and duck our heads at the sound of shells; but we do so without any conscious intervention of mind reasoning from cause to effect. We do it because we have inherited the reflex which compels us to act thus in response to the necessary stimuli.

It is often asked, If the action of instinct is thus so automatic and often so perfect in nature, where does the higher function of mind come in? Let me give an example in reply. I had in my house a wasps' nest in being. Three-fourths of the work of the insects during the season has been directed toward raising the large crop of queens and males which marks the end of the year. Every instinct of the nest has been for months adjusted to this social need. Yet what is the final result? The number of young queens in my nest is about 3,000, there being almost as many males. As the number of wasps in the world does not presumably increase, and as such a nest is always begun in the spring by a single queen, it follows that for one male and fertile female to attain their perfect end, some 5998 must on the average perish and fail. Such is the stupendous cost of life before the epoch of mind.

It is for reasons like these that there is to be observed everywhere throughout life one definite upward line of development, namely, the rising curve which marks the ascent of mind. We marvel at the complexity and history of the single cell in which the individual life in the higher forms always begins, a speck of matter capable of transmitting all the features and potentialities of inheritance which separate the various forms of life and distinguish one individual from another. But who can estimate the almost inconceivable complexity of the inherited forces which organize, in a single lifetime, the few ounces of grey matter of the human brain? But yesterday the components were a handful of inert material, to-day they have become the physical basis of the supreme reasoning consciousness of man and all that it implies. It is only by the comparative study and analysis of animal instinct and animal intelligence that we begin to have some feeble idea of the cost of the process in evolution, and of the unfathomable epochs of development which separate such a result from the first beginnings of life. No one who has grasped in any real sense the significance of mind in the evolution of life can hold the belief that the cycle of the manifestations of it which we have begun to witness will ever cease, or that it is destined to be in any way bounded even by the life of the planet on which we live.

XIII

THE BIRDS OF LONDON

THE rooks no longer build their nests in the Temple Gardens, and the thrushes and redbreasts, which, even fifty years ago, were wont to haunt the suburban gardens in the neighbourhood of what are now the main arteries of London traffic have long since retired before the ever-rising tide of bricks and mortar. Nevertheless what is left of London bird-life has not ceased to be interesting. On the contrary as the fog-pall has thickened over modern Babylon it has acquired a new interest which is peculiar to itself.

It is early morning in the month of May, and I am leaning against the window casement. It is light, but still some time before sunrise, and the air has that feeling which is peculiar to London air only in the spring in this hour out of twenty-four. The faint fresh odour brings into the mind for a moment a vision of a far off lake amongst my native hills from whose still surface the mist is just now beginning to rise, and the familiar cry of the coot as she sails out from the sedge, where during the night she has added another speckled egg to the store in her floating nest among the tall bulrushes. As I lean out of the window and catch the rumble of a belated cab my ears are filled with a peculiar noise which Londoners do not often listen to; for it is only to be heard about this time, and this is just the hour at which the great city falls into such short and fitful sleep as she gets. In the still air it sounds not unlike an army of stone-cutters at work with chisels and mallets on hard stone; but strange to say it does not come from anything so harsh as steel and stone, but from the throats of innumerable sparrows.

It is everywhere, along the street, on the slates overhead, in the trees in the gardens below, and a good deal of it comes from the sooty ivy on the wall where the birds have their nests. As the grey light grows brighter the eye begins to follow the movements of the birds in the back gardens below, and the sight is one worth seeing. It is the London sparrow at work in the breeding season during the first hour after the dawn. The incessant chirruping which goes on comes principally from the young birds. Some of them are still fledglings in the nests hidden away out of sight; others are standing about in lines and groups, along the ledge under the roofs, on the walls and palings, and on the branches of the trees. They are cold after the night and sit huddled up in their feathers, and they are all hungry. Their impatient cries drive the old birds frantic; I can see these going and coming in short quick flights over the opposite house to and from the deserted cab-rank in the adjacent street; they are hopping with quick anxious gait over the gravel below exploring everywhere for food; they are round the doors, on the window-sills and in the dust-bins. Few morsels will escape their sharp eyes; the city is asleep and they have the world to themselves.

An interesting study in bird-life is the London sparrow now. All the birds are not looking for food. Some are collecting building materials and are making short flights backwards and forwards, returning with straws, bits of rag, and odds and ends in their beaks. This is not the first venture in housekeeping with these; they have already reared one brood this year, and now they have begun again, and they will rear another before the season is out. The London sparrow is a by-word and proverb among birds for his breeding propensities; poor little fellow! it is the only way in which he can manage to make headway against the risks which continually beset his life, and the consequent high death-rate amongst his tribe.

Look at the crowd of eager nest-builders around that heap of house-sweepings against the dust-bin yonder. One after another of the little odds and ends of rubbish are taken up, weighed in the tiny bills, and found wanting according to some occult standard of the sparrow mind, until at last one suggests some element of fitness and the owner flies merrily away with his find. To give them their due these nest-builders look a somewhat disreputable lot. Sooty they are, hard worked, and with many a feather missing. The cab-horse has a luxurious and well-to-do look compared with a London sparrow in the height of the breeding season. The latter quarrels with his comrades for straws, loses his tail-feathers in duels and love-affairs, plucks out his breast feathers himself to line his nest, and works himself to the bone for his family in the intervals of quarrelling and love-making.

A quick harsh note and a flutter of wings. Every sparrow has left the ground. One looks round to find the cause of the alarm, but sees nothing at first. But we have been on the brink of a tragedy. A familiar form comes out from behind the wooden paling which tops the brick wall of the garden; it is my own cat, and he slinks into the open with that foolish sullen look peculiar to all the members of the feline tribe when they have been baulked of their prey. I call him softly by his name and he looks up and blinks his grey eyes at me. The marks of nocturnal dissipation are upon him. As he walks along the wall one may see the advantage of that grey fur striped with dark lines which is so common among the London cats; in the half light he is almost invisible on the dull background. The London cats mostly go their own ways and natural selection is only slightly tempered by human interference. This one walked into our house as a kitten and we took him in; he was housed and fed and petted; but a street arab he was born and will remain. From an early age he took to sparrow-hunting; we tried to break the old Adam in him, but after he had tasted blood and the pleasures of the chase the attempt had to be given up in despair. Some one sat in the room with him and a young tame sparrow for four hours, scarcely taking eyes off him. Blandishments were tried, but he was deaf to them; the attempt was given up and a stick was tried, but his spirit was unaffected. He feared the stick but he meant to have the sparrow—and he had it, under our eyes. He killed it with a stroke of his paw at a distance of some two feet even while he crouched down in fear from the punishment he knew would follow. I do not think any power could curb the lust for sparrow-killing in that grey blinking creature on the wall.

He is off now after some other mischief and the sparrows came back again. Along the flower-border there is a dark discoloured patch. It has been raining recently and it was here that the water collected in a shallow pool. The water is gone, absorbed by the sandy sub-soil beneath, and the surface is covered with a thin film of black mud, on which here and there the blades of a tiny bunch of grass lie stretched out, whitened now with the heavy dew they have gathered in the night. It is just the spot the earthworms like to come to the surface to feed in, and last night has been a night such as they love ; one can see the fresh casts which have been thrown up since the rain. One of the blue and pink burrowers has evidently come to the surface to stay, and he wriggles feebly and aimlessly on the moist ground. Presently a sparrow hops this way, the early bird is about to have his worm, you think. But no, he passes by and almost over it without appearing to see it.

The sparrow is no lover of creeping things, but it comes quite as a surprise to many of his admirers to learn that he is a vegetarian. Yet this is the trait in his character which will probably earn for him a place in history. It is because he is a vegetarian that the English sparrow has followed in the wake of the great Anglo-Saxon invasion of the world's wildernesses, even as his ancestors probably followed long ages ago in the wake of the Aryan invasion of Europe. The sparrow does not love the wood and the silent haunts of nature. He follows the settler with a very practical purpose in his head ; he comes to steal his corn, and to hang about the homestead to pick up scraps. He is no solitary hunter of winged and creeping things in waste places, but has always grown fat amongst the sheaves and pig-troughs of his patron. Nor has the revolution in our habits affected the sparrow. In these days some of us, alas ! no longer keep flocks and herds or grow our own corn ; we show an unmistakable tendency to crowd together in towns ; we shut out most of the sky and cover the face of nature for league upon league with bricks and asphalt ; nearly every feathered thing retires before the desolation we make. But the sparrow remains, for our habits suit him better than ever.

It is because the sparrow is a vegetarian that he is the only wild bird which really lives in London. We have many occasional feathered visitors to favoured spots in London, but none of them except the sparrow can truly be said to inhabit the great circle twelve miles in diameter which stretches outwards from St. Paul's. Here it is that the sparrow has the world practically to himself. For him our hundreds of miles of streets spread daily a bounteous feast ; even the poorest neighbourhoods find him a congenial home, and their dust-bins and cab-ranks spread a table continually before him in the presence of his enemies the cats. No wonder the London sparrow endures the soot and risks the cats ; few others of the feathered tribe have their daily bread provided so regularly.

It used to be said that the London sparrows went

out of town in August and took to the corn-fields. Some of the sparrows in the outskirts of the city may do this, but it cannot be true of the London sparrow proper, for he has no reason to migrate, and he is certainly never absent from his usual haunts. Did the London sparrow take it into his head to strike wing for the country it would be a vast exodus and the Kentish farmer might almost as hopefully prepare for a flight of locusts.

The song thrush and the blackbird are still visitors to the open spaces and private gardens in suburban London. The thrush may occasionally be both seen and heard in Kensington Gardens and Regent's Park, especially in the early morning. The thrush, though a shy bird, loves the earthworm, and he likes to hunt it amongst the short grass or under the fallen leaves, one reason doubtless why he still finds so many spots which suit him in and about London. It would be hard to find earthworms anywhere so plentiful as they are in many of the open spaces in London. Whether this is the result of abundant food and a favourable soil, or of the absence of the enemies which keep them in check, or of the great age of the turf, which is not broken up from time to time as it would be if under cultivation, it is difficult to say. Probably all three conditions have something to do with it. Kensington Gardens in particular is at the present time a splendid hunting ground; all through last winter, even in frosty weather, I was able to get a constant supply there for some frogs with no further aid than the point of my umbrella.

The starling is another bird which hunts the earthworm and which is occasionally to be seen on the turf in the Parks and open spaces in London. There is no bird which goes to work in such business-like fashion; his constant swingings from side to side so as to work the ground on both sides of him, the incessant jerking of his head up and down as he drives his beak inquiringly into the earth, and his motions varied every now and then by a short quick run as he seeks a more favourable spot, all combine to give one the idea that the bird feels he has not got a moment to lose over his work. The starling breeds in large numbers round London and is said to be on the increase in some neighbourhoods, Chislehurst for instance. He frequents the better class villa-residences a good deal, and likes to build in holes in trees or about houses. He particularly affects a hole in the wall out of reach or a broken roof. Starlings are generally to be seen in the open spaces in London in flocks of three or four birds to a dozen. One January I counted twenty-five birds in a single flock on the turf in Gray's Inn Gardens.

One of the most interesting birds which still figure in London bird life is, beyond doubt, the rook. His connexion with London is historic. We are all familiar with Goldsmith's experiences of the rooks which he watched at work on their nests in the Temple Gardens. The rook has however long since forsaken the precincts of the Temple and even living memory cannot now connect him with the place. But it may surprise many Londoners to hear that we have still a rookery in the very centre of London, a sight which certainly constitutes one of the greatest curiosities connected with the city.

Almost within a stone's throw of the heart of London, a little to the east of where Chancery Lane debouches into High Holborn, one may notice on the opposite side of the way a low archway. Through it a passage leads between high buildings to an open space nearly surrounded on all sides by legal offices. The place is known as Gray's Inn Gardens, and is well kept and little frequented. The sooty stretch of grass which looks as green and fresh as it is possible to look in the centre of London, is studded with a large number of tall plane trees in good condition which give the place a charmingly rural aspect quite unexpected in such a quarter. It is here, separated by some miles on every side from the open country, that there still exists in dwindling numbers one of the most ancient colonies of rooks; the nests still hang in the branches of the plane trees and up to the present the birds have always returned in the spring to put them in repair and hatch out their young.

At one time this rookery was far more extensive than it is now. Even in 1878 there were twenty-eight full nests in the breeding season; this year I count eighteen nests only. An interesting feature of the place, and one which, doubtless, tends to attach the colony to it, is the care which is taken of the birds. They are fed regularly, the food given being dog-biscuit steeped in water. It is spread by the gardener on an enclosed mound in the centre of the gardens, and it proves very attractive to a host of sparrows as well as to the rooks.

The rook, most conservative of all birds as he is, is now almost driven out of London. Even twelve years ago there were still several extensive rookeries in London. Writing so recently as 1878 Dr. E. Hamilton gives in the *Zoologist* an account of the rook in London which seems to separate the time by a long interval from the present. The rookery in Kensington Gardens was then still in existence and was said to contain thirty-one nests, which makes the writer recall with regret the year 1836 when the rookery extended from the Broad Walk to the Serpentine and contained close on one hundred nests. Since some of the higher trees were cut down in the gardens some years ago the birds have left the gardens, doubtless never to return, and there is not now a single nest in the place. Dr. Hamilton also mentions other places which the rooks then frequented but which they have since forsaken. He says: "In 1875 a rook's nest was built and the young hatched out in a tree at the back of Hereford Square, Brompton. The following year the birds returned with others and ten nests were built in the fine elm and plane trees there." But in 1879 there is a note in the same paper stating that the rooks' nests near Hereford Square, Brompton, which had been for several years frequented in the spring, had been that year deserted, the result being attributed to the noise of the workmen in the numerous buildings which were being erected in the vicinity.

This or a similar fate has now befallen nearly all the rook settlements in London. That the birds cling so long to their old haunts, despite many incongruous surroundings, is due to the well-known conservative instincts of the family. The rook is like the salmon: when he grows up he goes abroad far afield to sow his wild oats and seek his fortune, but when he settles down in life and elects to take upon himself parental responsibilities he always returns to the haunts of his youth. So it is that the family breeding grounds are tenanted from generation to generation until it becomes impossible to hold them any longer. Richard Jefferies once suggested the planting of the Thames Embankment thickly with trees in the hope of attracting the rooks to build there; but it is much to be doubted if this plan would now be successful; such feeding grounds as are within reach in London are now very restricted, and are much too frequented for the rook's taste.

The rook is however still occasionally to be seen in London. He used to affect the grounds of Lambeth Palace as much as anywhere, probably because of the seclusion. He might sometimes be seen there at work on the sward, or perched on a sooty branch of one of the trees that have become almost as black as his own plumage. In his visits to town he may be seen at times accompanied by his friend the jackdaw. It would be interesting to know the grounds of the friendship which everywhere seems to prevail between the rooks and the jackdaws. In the winter time in the country a flight of rooks is usually seen thickly interspersed with jackdaws. Starlings and other gregarious birds often fly with rooks too and mingle with them on the ground, but when they take to the wing the former always keep together. The jackdaws however mingle with the rooks indiscriminately both on the ground and on the wing and even in the roosting places. White of Selborne suggested that perhaps the jackdaws followed the rooks from interested motives: "because rooks have a more discerning scent than their attendants and can lead them to spots more productive of food. Anatomists," he quaintly adds, " say that rooks, by reason of two large nerves which run down between the eyes into the upper mandible, have a more delicate feeling in their beaks than other round-billed birds, and can grope for their meat when out of sight. Perhaps then their associates attend on them from motives of interest, as greyhounds wait on the motions of their finders, and as lions are said to do on the yelpings of jackals."

The jackdaws, like the rooks, used to be much commoner about London than they are now. They go in flocks in the winter but pair off in the breeding season. If they bred in London they would probably keep the sparrow down, for the jackdaw is rather an awkward neighbour for the smaller birds ; he robs their nests and carries off the unfledged young as dainty morsels. Church steeples and ivy-covered ruins within easy reach of the open country are the jackdaw's favourite breeding places. In the absence of such he has forsaken London at present ; but he will doubtless return to await the advent of Macaulay's New Zealander, for the promised sketch of the ruins of St. Paul's would not be complete without him. Cathedral towns he is generally associated with. The birds also build in the disused chimneys and continue dropping the twigs down until one lodges crosswise and holds the others, so enabling the foundations of the nest to be laid.

Although the rooks have forsaken Kensington Gardens some interesting country birds have recently established themselves there. In recent years some wood-pigeons have built their nests and reared their young in the Gardens, and these extremely shy birds may now be seen almost any day flying from tree to tree or on the ground feeding. These birds must not be confused with the true London pigeon of the blue-rock blood, which never takes to the trees and from which the wood-pigeon is quite distinct.

The term wild bird would technically exclude what is perhaps the most truly London bird after the sparrow, namely the pigeon, without which no description of bird-life in London would be complete. The London pigeon may not be called a wild bird but he is so in reality. He makes his nest where he pleases, and like the sparrow and the street arab, he lives in the streets. One of these days the London County Council may claim suzerainty over him; at present he owns allegiance to no man. Nearly all the larger public buildings and many of the churches in London are inhabited by pigeons; the birds make their nests in the inaccessible nooks and corners of the roofs and they increase and multiply from year to year. St. Paul's Cathedral, the British Museum, the Houses of Parliament, Somerset House, the Guildhall, the Law Courts, and nearly every building of the kind, has each its own particular flight of pigeons. These places with their carved masonry and wide spacious roofs with many an aerial nook and cranny offer just the kind of retreat which every descendant of the rock-pigeon loves. The pigeons which frequent some of the buildings are fed regularly, others forage for themselves, and it is one of the pleasantest sights of the city, and not an uncommon one, to see the London cabby emptying the remains of his nosebag in the middle of a flock of pigeons which show every sign of appreciation of the largess.

One of the most interesting things about the London pigeon is the way in which he is working out and confirming one of the most striking of the Darwinian theories. The wild pigeons in London are beyond doubt the descendants of stray birds which, finding food plentiful, took to their present mode of life, and their numbers are still occasionally recruited by tame birds which join them with the usual instinct of pigeons in such cases. The present pigeons are in fact the descendants of a motley crew of birds of many breeds and all colours. It is generally acknowledged that all varieties of our domestic pigeon came originally from one wild species, the common blue-rock, still found wild on many parts of the coast. This bird has a characteristic colour and very peculiar markings which distinguish it from all other species of pigeons throughout the world. The colour is slaty-blue, and the wings are marked with two dark transverse bands, the tail feathers having also a dark band across the end, while the outer tail-feathers are edged with white at the base. Despite the many distinct breeds of domestic pigeons at the present day, not only is it held that they are all descended from a common stock, but it is asserted, that if all the varieties were turned loose and allowed to interbreed freely, their descendants would, in course of time, all once more return to this blue-rock type in which they all originated. The London pigeon is doing something to work out this experiment. Any one who watches a flock of the pigeons which frequent the buildings in London will certainly see amongst them traces of many breeds and will find nearly all the colours represented. The blue-rock is, however, the predominant type and there is little doubt that if uninterrupted it would be only a question of time till it extinguished all minor peculiarities.

One result of the crowding of buildings in the central parts of London is that winged insect life is driven away, and as it has failed the swallows have retreated to the suburban fringes of London. The swallow, like the rook, has no objection to town life in itself, but insect food must be abundant to enable it to thrive. Early last September, great numbers of swallows were to be seen in the Crystal Palace neighbourhood circling high up in the air previous to their annual flight. For some days previously they were to be noticed from the South Eastern Railway, between St. John's and Grove Park stations, perched together in groups on every available roosting-place and chattering loudly, as they always do in these yearly meetings.

The swallow breeds freely round London. It is indeed curious to see the attachment of this shy gentle bird to the places frequented by man and the buildings used by him. The rafters of a roomy shed is the place which, above all others, the swallow loves to build in; failing this he is content with a place under the eaves or he will make shift as best he can with any other corner about the house. Like the sparrow the swallow has always been with us, and he probably twittered from his clay-built nest beneath the roof-tree of our Aryan forefathers; he has clung to us through all the varying phases of our architectural progress, and he takes to the capital of the Corinthian column as a nesting place as familiarly as he probably did to the crevices in the roof of the family cave in primeval times. Even our habit of living in towns does not drive him away, and it is only when his food supply fails that he retires from the London smoke and leaves us alone with the sparrow.

It may have occurred to others, as it has to me, to question whether some explanation is not to be sought of the curious habit which the swallow so persistently clings to, of building its nest about our houses. There is no reason why we should expect to find the swallow, like the sparrow, in association with man. It is by nature a shy bird; we do not provide for it in any way, for it subsists on a diet of insects which it hunts abroad on the wing; and, above all, it is a migrant, leaving us after a short interval for strange quarters in distant lands. Why is it that such a bird should come and build its nest familiarly round our windows and under our eaves? I have often wondered whether there may not be some connection between the instincts of the swallow and the rock-dwelling habits of our ancestors the cave-men. Judging by the relics which he has left behind him, primeval man must have occupied, and for enormously long periods, most of the suitable caves within reach in the greater part of the world. The swallow is naturally a cave-frequenting bird; it builds and breeds in great numbers about the roofs and walls of caves at the present time, and beyond doubt it must often have been the sharer of these rocky shelters of early man.

The sparrowhawk is a casual visitor to London and the neighbourhood, and like all his kind he is often mobbed by the swallows and other birds. Here on a southern common just outside the smoke zone one may see him sometimes. The swallows have been flying all the afternoon over the smooth surface of the pond, dipping occasionally into the tepid water, and in the still air sending the tiny wavelets travelling all the way to the distant edges. The house-martins, distinguished by the white patch on the lower part of the back, fly in and out amongst them. But what is this excitement which has suddenly come amongst the birds? They have forsaken the water and are flying overhead, the swallow's shrill excited note—tweet—tweet—coming from several throats at once. The eye travels inquiringly round. There is a flash of wings at the corner of the copse where the furze ceases and the white-thorns grow thickly, followed by a little bird-like cry of agony. A sparrowhawk has swooped down among the bushes and some little nest of half-fledged yellow-hammers hidden in the gorse has been orphaned. Now you may see the meaning of the swallow's note of alarm; the air is full of birds which seem to have gathered as if by magic. The hawk has secured his prey and stands for a moment holding it beneath him in his talons on a branch of the stunted oak. The swallows dash down furiously at him within an inch of his head, screaming loudly as they pass and rise again on the wing. He is off now with his prize in the direction of the wood, mobbed by the whole troop of birds which continue screaming in anger and making dashes at him the whole of the way. Nature is still red in tooth and claw even in these quiet neighbourhoods close to London. The excitement amongst the swallows does not calm down for a long time.

The great city grows apace and the feathered tribe retires steadily before it. Even our parks and open spaces do not seem to tempt the birds to linger with us. The nightingale still sings on Hampstead Heights, and the blackbird pipes on the fringes of Clapham Park; but even there they are in retreat before the speculative builder. Only the sparrow and the pigeon remain with us.

XIV

THE PLAGUE OF BIRDS

ONE of the results of the increased and growing interest in nature study in this country is a very curious one. Any one who has much experience of country habits, or who in particular has gained insight into the standards of the present generation of boys and girls in rural districts, must be struck with a change which has taken place. "The attitude of the country boy to birds and bird-nesting has much altered in my time," said recently an inspector of elementary schools of long and wide rural experience. A change is undoubtedly in progress; and the correlative is to be seen in the increased attention directed to nature subjects in many periodicals, the more general inculcation in elementary schools of a humanitarian attitude towards wild birds and their nests, and the gradual extension to nearly all parts of the country of the influence of the restrictive spirit of the Wild Birds' Protection Acts. This forms one aspect of the subject. To most thoughtful persons who have given attention to the matter these are the results which were to be expected and the change is regarded in a favourable light. Yet there is another side to the subject. It is now becoming clear that there has undoubtedly been of late years an enormous increase in the numbers of the commoner kinds of wild birds in this country. The protective causes just mentioned have operated very powerfully in favour of these birds. Their restrictive influences happen also to have been supplemented to a marked degree by the extension of game preserving which has taken place, and which has led in some districts to the wholesale extermination of the enemies of the birds or of their eggs and young in the breeding season. The result is forcing itself on attention in many places. It is a plague of birds which is attaining serious proportions.

In many of the home counties during the fruit season one wonders at the patience and endurance of the farmer and grower, so hard hit in many other ways, as one sees the extensive and organized service of precautions which has to be undertaken against the growing depredations of the birds. On the protection of the strawberry crop and the bush-fruit crop much labour and money have to be expended. It used to be the general custom to cover only wall fruit trees with nets, but it is now by no means uncommon to see the large trees and entire cherry orchards enveloped in a veil of netting. This means the sacrifice of one year's crop as an initial outlay. Without either this device or a constant service to scare the birds no fruit would be left on the trees. Starlings, blackbirds, and thrushes, the commonest of our country birds, are the principal offenders in this respect. All three kinds of birds have increased enormously in numbers in the South of England. Any intelligent observer who has gone bird-nesting in his youth may easily convince himself of this fact by noticing in the spring the prevalence of the nests of these birds nowadays as compared with former times. All these birds devour great quantities of both bush and tree fruit. Their appetites are enormous, and the rapidity with which they work is almost incredible. On a particular tree in a district in mid-Kent some late pears were allowed to hang last autumn until the end of September. During a few hours in which it was left unprotected a group of blackbirds managed to leave scarcely a single pear untouched of a bounteous crop, many of the larger fruit being eaten quite to the stem. It is a matter of common knowledge how greatly the sparrows have increased in many parts of England with the continuous growth of towns and the comparative safety these birds enjoy in urban neighbourhoods during the nesting season. In the autumn the sparrows come to fields some distance round London in almost incredible numbers just as the grain is ripening. The farmers say the birds come down from London for the season like the hoppers and bring their young ones with them. It must be heart-breaking to the cultivator of the land who has to live by the hard-won produce to see the havoc wrought by these birds. The writer a season or two ago walked along the edge of a seventy-acre wheat-field just beyond the Outer London district. There had been a scarcity of labour for a day or two in scaring the birds, and the sparrows had settled on the crop in vast crowds. He walked deep into the wheat in several directions, but was unable to find a single ear containing grain. It had all been picked out and nothing of value remained but the straw. All the excuses which used to be made for the sparrow as to the supposed services rendered to the agriculturist have now been exploded. He is known to be exclusively vegetarian, and so much a parasite on man and his labours that he is never found in woods or remote rural districts apart from human habitations. The rat is indeed a comparatively harmless creature compared with the common sparrow. In the country he will rob a wheat-field of its harvest in the manner described. In the suburban garden he will clear the rows of young peas as they appear above ground equally systematically. In the flower garden he will take the foliage of the pinks or carnations, or the blooms of the polyanthus, clearing the season's growth in a few days with the same businesslike thoroughness.

Another most destructive bird which has much increased in numbers in recent years in the South of England is the lesser blue-tit. This little bird is a great favourite in suburban gardens, where boxes are often put up for it to nest in, and where in the winter-time it is a common practice to hang out pieces of cocoa-nut for it to feast on. The havoc which this bird works is wrought in the winter-time in gardens and fruit plantations. Its favourite food at this season consists of the next season's buds of the red currant and gooseberry bushes. The damage which one little creature will work in a day is astonishing, and can hardly be credited by any one who has not actually seen it. The bird will alight on a twig of a gooseberry bush and clear every little rolled-up bud in which lies all the hidden promise of next season. It will rapidly go through the bush in this way. The birds come in family parties day after day to the same places, working systematically. The damage is scarcely perceptible to the naked eye until the following spring, when the bush is like a blind giant, full of sap but unable to grow through having all its eyes picked out. Many of them die in part or altogether, and numbers are permanently injured. Gardeners do not always put the injury down to the true cause, and bullfinches—most destructive birds in other ways and greatly on the increase—are sometimes blamed. A few months back the writer went over a fine old country place in one of the home counties, which had recently become vacant through the death of the owner. Noticing the well-known maimed look and the absence of promise of fruit on most of the fruit bushes, he spoke of it to the coachman, who accompanied him. "Yes," said he, with a grim smile, "it's them bullfinches. The old master wouldn't hear of touching 'em; but when the furniture went the gardener got a gun and the morning after shot over twenty of 'em."

There is no doubt that if the tillers of the soil were as vocal and had as much access to the periodical press as nature-lovers and bird-lovers, a very bitter cry would go up throughout the land against the increasing bird-plague and the damage that is is being done. The increase in game preserving has been mentioned as a secondary cause which operates by diminishing the number of the birds' natural enemies. It is no doubt a cause which has to be taken into consideration. The magpie, for instance, is a great destroyer of eggs and young birds. It used to be a fairly common bird in the

South of England a few generations ago. But over wide districts in Kent, Surrey and Sussex it has practically become extinct through the war waged against it. One may live for years in the country in these counties now without seeing one. The same may be said in lesser degree of the jay and some other bird enemies proscribed by gamekeepers. The balance of nature is tending to be disturbed.

XV

WHAT DO YOUNG ANIMALS KNOW?

IN these days when physicists are to be found discussing under new forms the old question as to the all-pervading character of mind in the universe, there is no subject which possesses more interest of a certain kind than that which relates to the mind of young animals. Any one who has made a systematic study of intelligence amongst young animals has generally found that his own mind has passed through various stages of growth. The belief in the simplicity of the subject soon gives way to a conviction of its profound complexity. It is an old Scottish tradition to incline to a certain reverence for the mind of the child, the view being that it possesses faculties and perceptions related to universal intelligence which are lost with later growth. It is a peculiarity of the study of the young that the observer often tends—even without admitting the fact to himself—to extend a somewhat similar view to the mind of young animals.

If a young queen wasp is imprisoned in the nest late in the autumn before she has mated, the sense of the latent future which is thwarted in her is a most striking spectacle to witness. That it is not simply the thwarting of the mating instinct which produces the results may readily be proved by experiment. The sense of the future which lies enfolded within her, and of the fact that her whole object in life is being defeated, is almost human to witness. When the same young queen wasp hibernates and survives the winter she does so undoubtedly through some unexplained effect on her nervous system of the knowledge of the fact that her part in the future is still to be played. For if she feels that she is being prevented from fulfilling it, she will die. When in the spring she seeks out the site for her underground nest, and proceeds single-handed to rear the first members of the future wasp-colony, every one of a long series of acts appears to the observer to be almost uncanny in their sequence, so clearly do they appear to be directed by an insight into needs in the future of which she can have no possible experience. It is this kind of knowledge in young animals which often conforms to Kant's definition of "pure reason" rather than intelligence of the ordinary kind which is so remarkable. If in the midst of all this preoccupied labour towards definite ends the same wasp be removed with her nest a few yards off, she will fly out and be quite unable to find it again. She will stupidly return time after time to the site where she first placed it. However imperative may be the indication to refer all the explanations of the working of the mind in young animals to the explanation of natural selection, we are still confronted with much that is inexplicable.

Take the example of a habit which is possessed by young wild ducks very soon after they are hatched out, and as they greedily seek their food in shallow muddy water. This habit is to stand up to the thighs in water, and, stamping gently and rapidly with their webbed feet on the muddy bottom, make the water rise in a constant eddy before them. It brings up any particles of food it may contain, which are then seized and devoured. The writer observed this spring young wild ducks hatched under a domestic hen practising this habit the third day after they had emerged from the egg. They every one stood and watched the muddy water eagerly as it swirled beneath their eyes, stamping rapidly on the bottom meanwhile and snatching continuously at the particles of food as they made them come to the surface. There was an evident knowledge of the action of the water under these complex movements which was quite surprising. No human actions could be better adjusted as a means to an end. The little ducks appeared by their movements and eager looks to know all about the cause and the effect as well as if they had been through the experience a thousand times. And yet they were but three days old! Some light is thrown on the subject by the actions of a young sheldrake, kept with the ducks, which went through nearly the same movements in searching for its food in the same water, but exhibited others quite as interesting. In its natural haunts the sheldrake feeds on the mud-flats and sands left by the receding tide. When it was fed on the dry ground it exhibited a very curious modification of the young wild ducks' habit. It went through a kind of dancing or prancing movement, stamping rapidly on the floor with its feet. The writer was interested to find, on looking the subject up, that

Darwin had noticed this habit in the sheldrake, and had attributed it to degraded reflex action—an instance, in short, of " an habitual and purposeless movement " in a changed environment. The sheldrake, it appears, in its natural haunts feeds on worms found in the sands and mud left uncovered by the tide. When a wormcast is discovered, it begins patting the ground with its feet, dancing as it were, over the hole, " and this makes the worm come to the surface." Hence the association of the stamping movement with its impatience for food on being fed. This may be ; but one would have thought better of the worm than that it should come out to be eaten on such an invitation. It might even be considered that natural selection would in its turn have developed the worm's intelligence so that it would have recognized so obvious a challenge from its enemy as an unmistakable danger-signal. Despite the great authority of Darwin, it seems reasonable to conclude that the young sheldrake's habit is related to its instinct or intelligence in stirring up the water so as to see its food in the muddy estuaries which form its natural haunts.

In the case of young animals all the old dispute as to where the border-line is to be drawn between reflex action and instinct, and again between instinct and reason, comes up, but with many new aspects. There is no doubt that young animals possess a remarkable kind of knowledge of the world which they afterwards lose to a great extent. It is often offered as one of the explanations of the unexplained problems of bird migration through vast distances that the young birds learn the way through travelling at first with a crowd of older birds which have done the journey before. In this way, it is pointed out, a tradition of the route would be passed down through indefinite generations of birds. There would be no mystery about the matter. The only difficulty about this explanation is that the young birds do not always go with the old. In the case of one of the greatest migratory species of all—namely, our common cuckoo—the young birds leave our shores many weeks later than the old ones. The conditions of life are so difficult for the cuckoo that the young have to remain in this way to attain their full strength and growth. How do the young find the way without any guidance or assistance from birds that have made the journey before? The answer sometimes given is that the young birds do not go at all, or at all events go only a short distance. There can be no doubt that this explanation is incorrect. Any one who has kept a young cuckoo through the autumn and winter months will feel convinced as to the immense distances which the young birds must traverse in flight during this period. For months during the declining season of the year every muscle of the young cuckoo's body will appear to be tense with the uncontrollable instinct of flight which seems to overmaster it. Preyer described the action of very young naked hermit-crabs, which at an early age have to find deserted shells in which to shelter themselves. Soon after leaving the egg they rush with extraordinary animation for suitable shells that are given to them in the water. They examine the opening at the mouth, and take up their quarters inside with remarkable alacrity. But if it chances that the shells are still occupied by molluscs, then they stay close by the opening and wait till the snail dies, which generally occurs soon after the beginning of the imprisonment and the strict watch. Upon this the small crab pulls out the carcase, devours it, and moves into the lodging himself. All this is said by other observers to be partly the result of inherited instinct and partly the result of the mind working in response to the call of urgent needs. It is doubtful, however, whether we know all that we think we do about the workings of mind in young animals in such circumstances. The more the subject is closely studied the less the observer finds himself inclined to accept ready explanations. The mind of young animals contains much that is very difficult to explain with our existing knowledge. Even in the case of the human child, all the present tendency of study is to show that it contains more than we recently would have found it possible to believe.

XVI

THE MIND OF A DOG

LONG ago Herbert Spencer set evolutionists thinking about the connection between intelligence in animals and the possession of a grasping organ. Parrots, squirrels, elephants, monkeys and many other animals were held to exemplify its existence. The explanation of the relationship has, of course, become obvious; for the creature which obtained the power of grasping could apply any intelligence it possessed so much more effectively than the same intelligence could be used by another animal without the power, that from the beginning natural selection doubtless placed a premium on the combination of the two faculties. In pursuit of this line of development, it has become an interesting fact that we are beginning to distinguish the existence of certain clues by which, as soon as we know them, we can understand much that would otherwise be obscure in the working of the intelligence of certain classes of animals. The intelligence of the dog is an example in point. When the mind of the dog is systematically compared with that of a monkey, it soon becomes evident that the former differs from the latter in quality.

The intelligence of the dog, that is to say, is quite different from that of the monkey in kind. Some time ago the writer was standing watching a monkey which was chained to a tree. The bystanders had been throwing him nuts. The monkey had eaten all within reach and had made several unsuccessful attempts to reach others which had fallen outside the radius of his chain. To the surprise and slight consternation of the little crowd watching him he snatched a stick from one of them and began deliberately to use it to rake the distant nuts within his reach. I have read how a dog, which had given proof of the highest intelligence in other experiments, was tried under somewhat similar circumstances. The animal was chained up and was given a stick, while a biscuit was put just out of his reach. When the biscuit was placed in the crook of the stick the dog rapidly learnt to pull in the stick with the biscuit. But he could never get beyond this point. The dog made no attempt to get the stick into the position in which he could use it. When a monkey, however, was tried in exactly the same circumstances he proved to have no difficulty at all in learning to obtain the biscuit by using a stick intelligently and almost in the manner of a human being. The inference usually drawn from facts of this kind is that the intelligence of the monkey is altogether superior to that of the dog. In short, it is this fundamental fact of the monkey's life, the perpetual handling of things, which gives us the clue as to the line along which the intelligence of the monkey has probably been evolved. We are led therefore to ask if there are similar clues by which we can better understand the intelligence of other animals. What, for instance, is the key to the inner workings of the mind of a dog?

Nearly all the probable ancestors of the dog are animals whose natural habit it is to hunt in packs. Wolves and jackals still do so. The Indian wild dogs and the hyæna dogs of South Africa exhibit similar habits. Even the scavenger dogs which infest the towns of the East show the same natural tendency to hunt in bands. We may take it, therefore, that the most fundamental instincts of the dog's mind have arisen out of association with his fellows for a common object like the hunting of game. At first sight the dog's more or less solitary life as the friend and associate of man would seem to take us out of the region of these ideas. Yet it will be seen on reflection that this is not so, and that it is probably in such ideas that we have now the clue to all the workings of the dog's mind and to the remarkable and exceptional kind of intelligence displayed by dogs in certain circumstances. One of the most obvious and striking of a dog's qualities is the sense of devotion and exclusive attachment to his master and his master's household. His eye will kindle at the approach of a friend. But not even the eye of the Oxford undergraduate who looked through Jude the Obscure without being even conscious of his presence could have been more unseeing than that of the dog can be when he looks through a stranger. His attitude is no doubt prompted by feelings towards an outsider whom he regards as not a member of his pack. Similarly as to the dog's extraordinary loyalty, the remarkable sense of obedience which will hold him to a command for days and weeks, the power of control which he exercises over his strongest emotions, and the innate moral sense which may be observed to render a well-trained dog miserable if he fails in what is expected of him. All these qualities are characteristic of the dog. They are of a very high order: and yet they would seem to be less highly developed in the monkey than in the dog. The clue to the dog's mind is probably that the ideas related to these qualities were originally connected with his place in association with others in pursuing or attaining a common object. The dog has probably still some sort of conception of his place as member of a co-operative group and of his master as the wise and resourceful leader of it. In those most remarkable displays of almost human intelligence on the part of dogs, where the power of instantly comprehending the nature and possibilities of a locality or the contents of a difficult situation is involved, we probably see the dog's mind at its best. For here the governing ideas are probably of the kind which originally had their function in intelligent co-operation in hunting with companions. In the various breeds of dogs we get these fundamental ideas carried by development in widely different directions. In the pointer and setter the co-operation of the dog with his master to circumvent the game still remains obvious, even under highly artificial conditions. In many other characteristic qualities of the dog's mind, as displayed in various breeds, we may obtain much light on the working of canine intelligence if we keep this clue to the dog's original nature always before us. To many persons, for instance, one of the most difficult cases to explain in the light of this hypothesis would be that of the collie or sheep-dog. The writer was recently staying on a South African farm where the owner was seriously plagued with the ravages amongst his sheep of black-backed jackals. The extraordinary intelligence of the animals in shepherding the prey to their malign purposes was bitterly remarked upon. As the outward resemblances of a jackal to a collie were pointed out, the question was asked as to how we could imagine any relationship between animals whose fundamental instincts appeared to be so widely apart. Thus the sheep-dog was the friend of man and its leading characteristic was a desire for the preservation of the sheep and the power to employ most remarkable instincts in furthering his master's purposes to this end. The jackals, on the contrary, regarded the sheep simply as their natural prey. Yet the explanation even in this case is probably not far to seek when we have the clue to a dog's mind. For in a dim way the ordinary collie probably regards the sheep as no more than property or game belonging to his pack. He thinks of himself in all probability as assisting the wise dog at the head of the pack in the exciting occupation of shepherding the captured game. That there is a close natural relationship between the jackal and the collie as regards the sheep one pregnant fact will illustrate. Every experienced shepherd knows that a collie is more liable than most other dogs to take to killing or worrying sheep, and that when this happens he is the most inveterate, the most cunning and the most to be feared of all dogs. The cause is easy to understand. The degeneration in the dog's mind in this case has probably followed the path of transition from the social duty of taking care of the game in the joint interest to the ultimate, but now illegitimate, purpose of killing and eating it.

XVII

INTELLIGENCE OF SQUIRRELS

TO our common squirrel belongs the distinction of being almost the only native creature we possess which furnishes an example of a curious fact remarked upon by Herbert Spencer. Throughout nature, he pointed out, there exists among all classes of animals a marked relationship between the power of grasping or handling and a high level of intelligence. It is this power of grasping—for it holds its food and all objects which attract its interest in almost human fashion—combined with the high intelligence it displays on almost all occasions, which renders the squirrel such a fascinating object of study. In March, when occasional sunny days begin to divide the cold spells of weather, the squirrel may often be seen to advantage on the bare boughs in the more secluded woods and copses. The little creatures still wear their winter outfit, the dark grey tints of the fur, so different from the rich, ruddy brown of the summer, harmonizing well with the naked trees and no doubt tending to make the owners less conspicuous in the absence of all cover. Squirrels, although they pass the greater part of the winter asleep, snugly rolled together in the warm nests which they construct for themselves, feed to a considerable extent even in the colder months. They wake up in the fine intervals and come out to visit the stores of nuts and acorns they have hidden away in the autumn.

Its winter habits seem to give the squirrel a half-way place between animals which during the cold season fall into a state of suspended animation almost resembling death and those which remain in the ordinary functional state of activity. It is very interesting to watch the onset of the hibernating mood, which appears to be accompanied by deep physiological changes. In a pair which the writer had under observation last autumn, in conditions closely resembling those in nature, the great activity which is natural to squirrels in the summer season began to fail as early as the end of August. The little creatures passed a longer and longer time asleep each day as the year declined, and towards the end of November they came out for only a few minutes daily to be fed. Still later they could not be roused during cold weather, and the animal functions seemed at last in large part suspended.

The intelligence of squirrels makes them often resemble miniature monkeys. Even in its wild state, in its gambols on the trees, our native squirrel always seems as if it were conscious of being watched and to have, like a monkey, a certain eye to effect in all its doings. It never appears, like most wild animals, to want simply to disappear from view, but will mostly manage to keep in sight of the spectator. There are curious lacunæ in the intelligence of squirrels when kept under observation which would seem to suggest that a very considerable mind development, comparatively measured, lies behind intelligent acts of animals higher in the scale. A squirrel will, for instance, almost from its birth hold with its hands the food which it eats. It will grasp and handle and use its hands and fingers in some respects almost as intelligently as a human being. Yet it retains till it is almost full grown a very curious limitation marking its relationship to lower and less intelligent forms of life. It always attempts, the writer has observed, to seize hold of things with its mouth, never with its hands. So ingrained is this remarkable peculiarity that for the first months of its existence the squirrel, although using its hands freely to hold things, will never think of using them to reach things. If it should be unable to seize the food offered to it with its mouth it will think it out of reach, and will go hungry, even though the food be in reach of hand or arm. It will never think of using these to obtain the food.

Squirrels bury stores of nuts and other food during the autumn, but they often entirely forget what they have done, or where they have placed their hoards. This habit of burying food seems to be not entirely a matter of intelligence with squirrels. It is in all probability largely an instinctive or automatic habit. The pair the writer had under observation would perform the make-believe of burying a nut in the floor of a room. They would press the nut down on the carpet and go through all the motions of patting the earth over it, after which they went away, apparently satisfied that the nut was safely buried. Many persons must have observed a habit of the little chipmunk squirrels, often kept tame in America. This squirrel when cleaning itself, after the manner of a cat—as squirrels of all kinds do, with charming and taking movements—is given to sneezing into its paws as if to damp them for application to its fur. It has become a moot question with observers as to how far this action is the result of intelligence or merely reflex. There can be no doubt, however, of the high level of intelligence amongst squirrels of all kinds, the true tree-squirrels being, as a rule, more gifted than those which live in the ground or burrow in the earth. Even the little ground chipmunks, which are so plentiful in Western America, suggest the unusual intelligence of the squirrel in every movement. When the writer was in South California he used silently to watch them playing on the ground and be struck by the same consciousness in their movements of being under observation which is so characteristic of our English squirrel in its antics in the trees overhead. It is said that these ground squirrels when kept in captivity will get as used to a revolving cage as the ordinary tree-squirrel, and will take just as much delight in making it turn rapidly.

It is an interesting fact, the cause of which remains at present unexplained, that our common squirrel is much on the increase in certain parts of Great Britain. Over wide districts in Scotland where the squirrel is now very numerous it was an unknown animal a few decades ago, and this though thick woods abounded. In many places observant proprietors attribute the present invasion of the squirrel to a cycle of milder seasons, but it would be difficult to say whether this is the correct explanation. Squirrels seem to be able to adapt themselves to all kinds of climates and weathers. The facts already mentioned of the hibernation of the squirrel in this country show how easily its habits might graduate, either towards activity all the winter through in a milder climate, or towards absolute torpor and insensibility throughout the whole season, as happens in North America and in other colder countries. It is probably this adaptability of habit, coupled with the fact that no kind of food, animal or vegetable, comes entirely amiss to it, that makes the squirrel so cosmopolitan. It is one of the most widely distributed of families. While the squirrel is nearly related to the rat and the rabbit, many of the burrowing forms closely imitating the habits of these creatures, it quite rises to the suggestion of the monkey in its arboreal habits and general intelligence. The flying squirrels have even attempted the solution of some of the problems of aerial navigation. Squirrels of all kinds breed rapidly. The month of March in this country marks the resumption of a life of full activity after the winter. They pair soon after they wake up and there are sometimes two broods, each of three or four or more, in the season. If taken young squirrels make delightful pets, their gambols and intelligent antics being a constant source of pleasure. They seem to be almost without natural fear of persons, readily treating them quite as companions and evidently regarding them as no more than wiser squirrels of a larger kind.